QUESTION & WORKBOOK

Cambridge International AS & A Level

Mathematics
Mechanics

Jean-Paul Muscat

Working for over 25 YEARS WITH Cambridge Assessment International Education

HODDER
EDUCATION

Answers to all the questions can be found at www.hoddereducation.com/cambridgeextras

Questions from Cambridge International AS & A Level Mathematics papers are reproduced by permission of Cambridge Assessment International Education. Cambridge Assessment International Education bears no responsibility for the example answers to questions taken from its past question papers which are contained in this publication.

This text has not been through the Cambridge International endorsement process.

Hachette UK's policy is to use papers that are natural, renewable and recyclable products and made from wood grown in sustainable forests. The logging and manufacturing processes are expected to conform to the environmental regulations of the country of origin.

Orders: please contact Bookpoint Ltd, 130 Park Drive, Milton Park, Abingdon, Oxon OX14 4SE. Telephone: +44 (0)1235 827827. Fax: +44 (0)1235 400401. Email education@bookpoint.co.uk Lines are open from 9 a.m. to 5 p.m., Monday to Saturday, with a 24-hour message answering service. You can also order through our website at www.hoddereducation.com.

ISBN: 978 1510 421837

© Jean-Paul Muscat 2018

Published by Hodder Education, an Hachette UK Company
Carmelite House, 50 Victoria Embankment
London EC4Y 0DZ

www.hoddereducation.com

Impression number   10 9 8 7 6 5 4 3 2

Year   2022 2021 2020 2019

Cover photo by Shutterstock/ArtisticPhoto

Illustrations by Integra Software Services Pvt. Ltd., Pondicherry, India

Typeset in Minion Pro 10.5/14 by Integra Software Services Pvt. Ltd., Pondicherry, India

Printed in the UK

A catalogue record for this title is available from the British Library.

Contents

Formulae

Uniformly accelerated motion

$$v = u + at, \quad s = \frac{1}{2}(u + v)t, \quad s = ut + \frac{1}{2}at^2, \quad v^2 = u^2 + 2as$$

Throughout the exercises you should use $g = 10 \, \text{m s}^{-2}$ for the acceleration due to gravity.

1 Motion in a straight line

1.1 The language of motion

1 For the following displacement–time graphs calculate the total overall displacement and the total distance travelled.

(i)

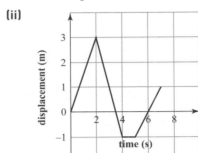

(ii)

2 A particle is moving from east to west in a straight horizontal line so that its position x at time t is given by
$x = 7 - t^2(4 - t)$.

   (i) What is the position of the particle at times $t = 0, 0.5, 1, 1.5, 2, 3, 4, 5$?

   (ii) What is the displacement from the original position after 5 s?

   (iii) At what time is the particle furthest east from its original position and what is its position then?

   (iv) What is the total distance travelled by the particle in the first five seconds?

3 A ball is thrown straight up in the air so that its displacement is given by $x = 2 + 12t - 5t^2$.

(i) Sketch a displacement–time graph for $0 \leqslant t \leqslant 2.5$.

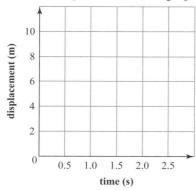

(ii) Find the displacement relative to its starting position after 2.5 s.

(iii) At what time is the highest point reached and how high does the ball go?

(iv) What is the total distance travelled in the 2.5 s?

4   A point P is moving in a vertical straight line. The displacement–time graph is shown.

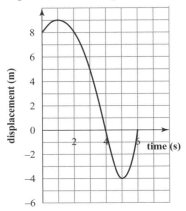

(i)   Find the greatest displacement of P above its initial position.

(ii) Find the largest distance of P from its initial position.

(iii) Find the time interval in which P is moving downwards.

(iv) Find the times when P is instantaneously at rest.

(v)   Find the total distance travelled.

1.2 Speed and velocity

1. For the displacement–time graphs (i)–(iv), find

   (a) the initial and final positions

   (b) the total displacement

   (c) the total distance travelled

   (d) the velocity and speed for each part of the journey

   (e) the average velocity for the whole journey

   (f) the average speed for the whole journey.

(i)

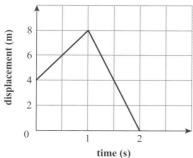

(ii)

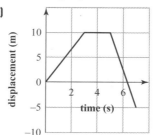

(iii)

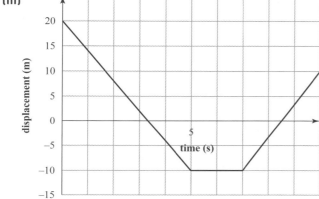

(iv)

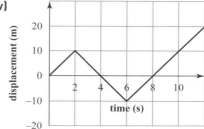

2. A bus leaves town A at 2:00 pm and travels to town B, 40 km away at a speed of 30 km h^{-1}.

   A car leaves A at 2:30 pm, travels towards B along the same road as the bus and travels at 80 km h^{-1}.

   (i) Sketch a displacement–time graph to show the motion of the two vehicles.

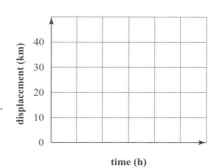

   (ii) Find at what time the car catches up with the bus.

   (iii) How far from A are they then?

3 A, B and C are three points on a straight road with AB = 1200 m, BC = 200 m and B is between A and C. A boy cycles from A to B at 10 m s^{-1}, pushes his bike from B to C at an average speed of 0.5 m s^{-1} and then cycles back from C to B at an average speed of 15 m s^{-1}.

(i) Find the average speed of the boy for the whole journey.

(ii) Find the average velocity of the boy for the whole journey.

4 A car travels 50 km from A to B at an average speed of 80 km h^{-1}. It stops at B for 30 minutes and then returns to A travelling at an average speed of 60 km h^{-1}.

(i) Find the total time taken for the whole journey.

(ii) Find the average speed for the whole journey.

(iii) Find the average velocity.

5 At what average speed do you need to run to travel a mile under 4 minutes? Give your answer in mph and in S.I. units. [1 mile = 1609.34 m]

1.3 Acceleration

1 From the velocity–time graph, find

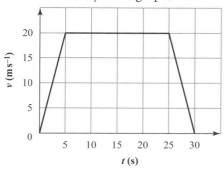

(i) the acceleration at times 2 s, 10 s and 30 s

(ii) the total distance travelled.

2 The figure shows an acceleration–time graph modelling the motion of a particle moving in a straight line.

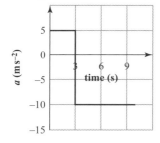

(i) At time $t = 0$ the particle has a velocity of $5\,\mathrm{m\,s}^{-1}$ in the positive direction.

Find the velocity of the particle when $t = 3$.

(ii) At what time is the particle travelling in the negative direction with a speed of $5\,\mathrm{m\,s}^{-1}$?

3 The velocity–time graph illustrated shows the motion of a particle in a straight line.

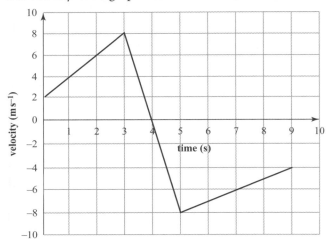

(i) Find the acceleration in the three stages of the motion, then use this to sketch an acceleration–time graph for the motion.

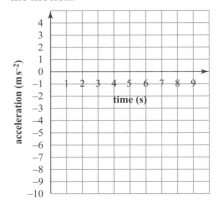

(ii) Find the displacement of the particle at times $t = 3$, $t = 4$, $t = 5$ and $t = 9$.

4 A particle travels along a straight line. Its acceleration in the interval $0 \leqslant t \leqslant 6$ is shown on the acceleration–time graph.

(i) Write down the acceleration at time $t = 2$.

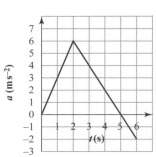

(ii) Given that the particle starts from rest at $t = 0$, find the speed at $t = 2$.

(iii) Find an expression for the acceleration as a function of t in the interval $0 \leqslant t \leqslant 2$.

(iv) At what time is the speed greatest?

(v) Find the change in speed from $t = 2$ to $t = 6$, indicating whether this is an increase or a decrease.

5 A particle starts from rest at time $t = 0$ and moves in a straight line accelerating as follows

$a = 2; \qquad 0 \leqslant t \leqslant 10$

$a = 0.5; \quad 10 < t \leqslant 50$

$a = -3; \quad 50 < t \leqslant 60$

where a is the acceleration in $\mathrm{m\,s^{-2}}$ and t is the time in seconds.

(i) Find the speed of the particle when $t = 10$, 50 and 60.

(ii) Sketch a speed–time graph for the particle in the interval $0 \leqslant t \leqslant 60$.

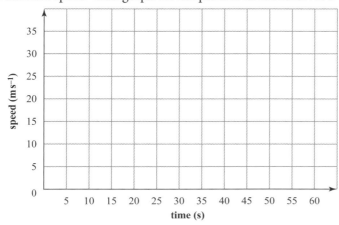

(iii) Find the total distance travelled by the particle in the interval $0 \leqslant t \leqslant 60$.

1.4 Using areas to find distances and displacements

1 Use the velocity–time graph.

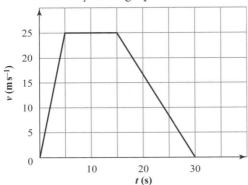

(i) Find the acceleration when $t = 20$.

(ii) Find the distance travelled in the first 10 seconds.

(iii) Find the total distance travelled.

2 A toy car is moving in a straight line. The velocity–time graph is shown.

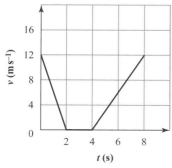

(i) Calculate the distance travelled by the car from $t = 0$ to $t = 8$.

(ii) How much less time would it have taken for the car to travel this distance if it had kept its original speed?

(iii) Find the acceleration at times $t = 1$ and $t = 8$.

(iv) From $t = 8$ to $t = 20$, the car travels a further $180\,\text{m}$ with a new uniform acceleration $a\,\text{m}\,\text{s}^{-2}$. Find the velocity at time $t = 20$ and hence find a.

3 A particle moves along a straight line. It starts from rest, accelerates at $3\,\text{m}\,\text{s}^{-2}$ for 2 seconds and then decelerates at a constant rate, coming to rest in a further 6 seconds.

(i) Sketch a velocity–time graph.

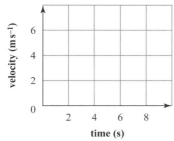

(ii) Find the total distance travelled.

(iii) Find the deceleration of the particle.

(iv) Find the average speed for the whole journey.

4   The figure shows the velocity–time graph for the motion of a particle in a straight line.

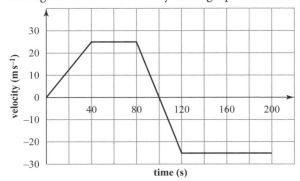

(i)   Find the displacement of the particle at times $t = 40$, $t = 80$, $t = 120$ and $t = 200$.

(ii) Find the total distance travelled.

(iii) At what time does the particle pass its starting point?

5 A train travels from A to B, a distance of 50 km. Starting from A it takes 5 minutes to accelerate uniformly to 40 m s^{-1}, maintaining this speed until, with uniform deceleration over 2.5 km, it comes to rest at B.

    (i) Find the distance travelled while accelerating.

    (ii) Find the time taken to decelerate.

    (iii) Find the time taken for the journey.

Further practice

1 A stone is catapulted vertically upwards so that its position y at time t is as shown on the graph.

    (i) Write the position of the stone at times 0, 0.5, 1, 1.5, 2, 2.5 and 3.

    (ii) Find the displacement of the stone relative to its starting position at these times.

    (iii) What is the total distance travelled

        (a) during the first 1.4 s
        (b) during the 3 s of the motion?

2 The position of a particle moving along a straight line is $x = 3t^2 - 14t + 11$ where x is in metres and t is in seconds with $0 < t < 5$.

    (i) At what times does $x = 0$?

    (ii) When is the particle furthest from its starting point?

    (iii) How far has the particle travelled in the first 5 seconds?

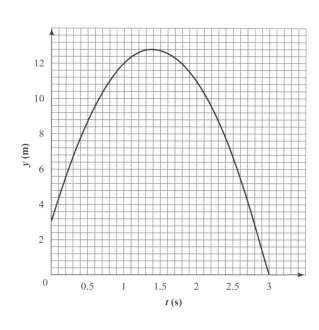

3 A particle is moving in a straight line so that its position x m at time t is as shown in the graph.

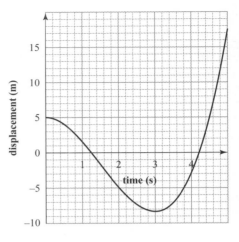

(i) Write the position of the particle at 0.5 s intervals from $t = 0$ to $t = 5$.

(ii) Find the displacement from its starting position of the particle at these times.

(iii) At what time is the particle furthest from its starting point?

(iv) Find the total distance travelled.

4 The world record for the men's 1 km time trial in cycling is 56.303 seconds. Find the average speed, giving your answer in $km\,h^{-1}$.

5 I first see lightning and then hear the sound of thunder 3 seconds later. Find how far away the lightning struck. [Assume the speed of sound is $340\,m\,s^{-1}$.]

6 A car travels 50 km from A to B at an average speed of $100\,km\,h^{-1}$. It stops at B for 45 minutes and then returns to A. The average speed for the whole journey is $40\,km\,h^{-1}$.

(i) Find the average speed from B to A.

(ii) Find the average velocity for the whole journey.

7 A car travels along a straight road. The car starts at rest from the point A and accelerates for 30 s at a constant rate until it reaches a speed of $20\,m\,s^{-1}$. The car continues at this speed for T s and then decelerates at a constant rate for 20 s until the car slows down to $10\,m\,s^{-1}$ as it passes the point B. The distance AB is 5 km.

(i) Sketch a velocity–time graph for the journey between A and B.

(ii) Find the total time taken for the journey.

8 The motion of a coach travelling along a straight road follows three stages. Initially travelling at $18\,m\,s^{-1}$, the coach decelerates uniformly for 10 s, reaching a velocity of $12\,m\,s^{-1}$. During the second stage of the motion, the coach travels at a constant speed of $12\,m\,s^{-1}$ for 20 s. In the third stage, the coach accelerates uniformly, reaching a velocity of $18\,m\,s^{-1}$ after a further 20 s.

(i) Sketch the velocity–time graph for the motion.

(ii) Find the average speed of the coach.

9 The velocity–time graph shows the motion of a particle along a straight line.

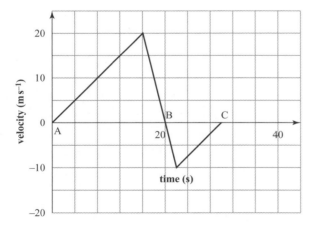

(i) The particle starts at A at $t = 0$ and moves to B in the next 20 s. Find the distance AB.

(ii) T seconds after leaving A the particle is at C, a distance 50 m from B. Find T.

(iii) Find the displacement of C from A.

10 A cyclist can cycle uphill at an average speed of $15\,\mathrm{km\,h^{-1}}$, downhill at an average speed of $60\,\mathrm{km\,h^{-1}}$ and on the flat at $45\,\mathrm{km\,h^{-1}}$. Find the cyclist's average speed on each of the following stages of a race.

(i) A mountain stage with 80 km on the flat, 50 km uphill and 30 km downhill.

(ii) A flat stage consisting of 150 km on the flat, 5 km uphill and 10 km downhill.

11 A train has a maximum allowed speed of $20\,\mathrm{m\,s^{-1}}$. With its brakes fully applied, it has a deceleration of $0.5\,\mathrm{m\,s^{-2}}$. If it can accelerate at a constant rate of $0.25\,\mathrm{m\,s^{-2}}$ find the shortest time in which it can travel from rest in one station to rest in the next station 10 km away.

12 The maximum acceleration for a lift is $2\,\mathrm{m\,s^{-2}}$ and the maximum deceleration is $5\,\mathrm{m\,s^{-2}}$. Find the minimum time for a journey of 40 m:

(i) if the maximum speed is $6\,\mathrm{m\,s^{-1}}$ (ii) if there is no restriction on the speed.

13 A cyclist starts from rest and takes 5 seconds to accelerate at a constant rate up to a speed of $12\,\mathrm{m\,s^{-1}}$. After travelling at this speed for 50 s, the cyclist decelerates to rest at a constant rate over the next 5 s.

(i) Sketch the velocity–time graph.   (ii) Find the distance travelled.

14 The graph shows how the velocity of a particle varies in a 30 second period as it moves in a straight line.

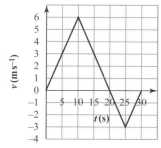

(i) At what time is the particle furthest from its starting point and how far is it then?

(ii) Find the displacement of the particle at the end of the 30 seconds.

(iii) Find the total distance travelled.

15 A particle is travelling in a straight line. Its velocity $v\,\mathrm{m\,s^{-1}}$ at time $t\,\mathrm{s}$ is given by $v = 5 + 3t$ and $0 \leqslant t \leqslant 4$.

(i) Write down the initial velocity and the acceleration.

(ii) Write down the velocity of the particle when $t = 4$ and find the distance travelled in the first 4 seconds.

For $4 \leqslant t \leqslant 12$ the acceleration of the particle is $-2\,\mathrm{m\,s^{-2}}$.

(iii) Write down an expression for the velocity in this interval.

(iv) Find the total distance travelled by the particle in the first 12 seconds.

16 A cage goes down a vertical mine shaft 425 m deep in 48 s. During the first 14 s it is accelerated from rest to its maximum speed. For the next 20 s it moves at this speed. It is then uniformly decelerated to rest. Find the maximum speed it attains.

17 A train has a maximum allowed speed of $25\,\mathrm{m\,s^{-1}}$. With its brakes fully applied, it has a deceleration of $1\,\mathrm{m\,s^{-2}}$. If it can accelerate at a constant rate of $0.5\,\mathrm{m\,s^{-2}}$ find the shortest time in which it can travel from rest in one station to rest in the next station 8 km away.

18 The driver of a train travelling at $150\,\mathrm{km\,h^{-1}}$ on a straight level track sees a signal to stop 1 km ahead and, putting the brakes on fully, comes to rest at the signal. He stops for 2 minutes and then resumes the journey, attaining the original speed in a distance of 5 km. Assuming constant acceleration and deceleration, find how much time has been lost owing to the stoppage.

19 A train starts from rest at A and accelerates uniformly at $0.5\,\mathrm{m\,s^{-2}}$ for 1 minute. It then travels at a constant speed for 20 minutes, after which it is brought to rest at B with a constant deceleration of $2\,\mathrm{m\,s^{-2}}$. Sketch a velocity–time graph and use it to find the distance AB.

20 The velocity–time graph for a particle P moving in a straight line AB is shown.

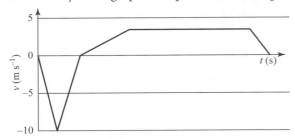

The particle starts at rest from a point X, $\frac{1}{3}$ of the way between A and B and moves towards A. It comes to rest at A when $t = 5$ s.

(i) Find the distance AX.

In the second stage of the motion, P is starting from rest at $t = 5$ and moves towards B. The particle takes 45 seconds to travel from A to B and comes to rest at B. For the first 10 seconds of this stage it is accelerating at $0.25\,\text{m s}^{-2}$, reaching a velocity V which it keeps for T seconds after which it decelerates to rest at B.

(ii) Find V.

(iii) Find T.

(iv) Find the deceleration.

Past exam questions

1

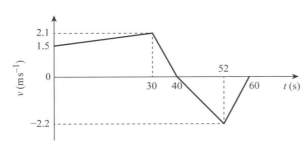

A woman walks in a straight line. The woman's velocity t seconds after passing through a fixed point A on the line is $v\,\text{m s}^{-1}$. The graph of v against t consists of 4 straight line segments (see diagram). The woman is at the point B when $t = 60$. Find

(i) the woman's acceleration for $0 < t < 30$ and for $30 < t < 40$, [3]

(ii) the distance AB, [2]

(iii) the total distance walked by the woman. [1]

Cambridge International AS and A Level Mathematics 9709 Paper 43 Q1 November 2011

2 A particle P moves in a straight line. It starts from rest at point O and moves towards a point A on the line. During the first 8 seconds P's speed increases to $8\,\text{m s}^{-1}$ with constant acceleration. During the next 12 seconds P's speed decreases to $2\,\text{m s}^{-1}$ with constant deceleration. P then moves with constant acceleration for 6 seconds, reaching A with speed $6.5\,\text{m s}^{-1}$.

(i) Sketch the velocity–time graph for P's motion. [2]

The displacement of P from O, at time t seconds after P leaves O, is s metres.

(ii) Shade the region of the velocity–time graph representing s for a value of t where $20 \leqslant t \leqslant 26$. [1]

(iii) Show that, for $20 \leqslant t \leqslant 26$, $s = 0.375t^2 - 13t + 202$. [6]

Cambridge International AS and A Level Mathematics 9709 Paper 42 Q6 June 2013

3 A car travels in a straight line from A to B, a distance of 12 km, taking 552 seconds. The car starts from rest at A and accelerates for T_1 s at $0.3\,\text{m s}^{-2}$, reaching a speed of $V\,\text{m s}^{-1}$. The car then continues to move at $V\,\text{m s}^{-1}$ for T_2 s. It then decelerates for T_3 s at $1\,\text{m s}^{-2}$, coming to rest at B.

   (i) Sketch the velocity–time graph for the motion and express T_1 and T_3 in terms of V. [3]

   (ii) Express the total distance travelled in terms of V and show that $13V^2 - 3312V + 72\,000 = 0$. Hence find the value of V. [5]

Cambridge International AS and A Level Mathematics 9709 Paper 43 Q5 November 2013

4 A train starts from rest at a station A and travels in a straight line to B, where it comes to rest. The train moves with constant acceleration $0.025\,\text{m s}^{-2}$ for the first 600 s, with constant speed for the next 2600 s, and finally with constant deceleration $0.0375\,\text{m s}^{-2}$.

   (i) Find the total time taken for the train to travel from A to B. [4]

   (ii) Sketch the velocity–time graph for the journey and find the distance AB. [3]

   (iii) The speed of the train t seconds after leaving A is $7.5\,\text{m s}^{-1}$. State the possible values of t. [1]

Cambridge International AS and A Level Mathematics 9709 Paper 41 Q5 June 2011

STRETCH AND CHALLENGE

1 A car travels a distance of 20 km from A to B taking 16 minutes. The journey is in three stages. In the first stage the car starts from rest at A and accelerates uniformly to reach a speed of $V\,\text{m s}^{-1}$. In the second stage the car travels at constant speed V for 15 minutes. During the third stage of the journey the car decelerates uniformly coming to rest at B.

   (i) Sketch the velocity–time graph for the journey.

   (ii) Find the value of V.

   (iii) Given that the ratio of acceleration : deceleration is 1 : 2, find how long it takes to accelerate in the first stage of the motion and the distance travelled while the car decelerated in the third stage of the journey.

2 A lift makes the first part of its descent with uniform acceleration a and the remainder with uniform deceleration $2a$. Prove that, if D is the distance travelled and T is the time taken, then $D = \frac{1}{3}aT^2$.

3 A particle moves in a straight line. It starts from rest at A and moves from A to B with constant acceleration $3a$. It then moves from B to C with acceleration a and reaches C with speed V. The times taken in the motion from A to B and from B to C are equal to T. Find T in terms of V and a. Show that the ratio of the distances AB : BC is 3 : 7.

4 Two stations A and B are 2 km apart on a straight track. A train starts from rest at A and comes to rest at B. The train accelerates uniformly for $\frac{3}{4}$ of the distance and decelerates uniformly for the remainder. The journey takes 4 minutes. Find

   (i) the acceleration

   (ii) the deceleration

   (iii) the maximum speed of the train.

2 The constant acceleration formulae

1 The following questions involve motion under constant acceleration. All lengths are in metres, time in seconds.

(i) Find v when $u = 7$, $a = 2$ and $t = 3$.

(ii) Find s when $u = 5$, $v = 10$ and $t = 5$.

(iii) Find s when $u = 2$, $a = -2$ and $t = 4$.

(iv) Find u when $a = -4$, $s = 2$ and $v = 3$.

2 A cyclist accelerates uniformly along a straight horizontal road so that when she has travelled 20 metres, her velocity has increased from $8 \, \text{m s}^{-1}$ to $10 \, \text{m s}^{-1}$. Find the acceleration of the cyclist and the time it takes her to travel the distance.

3   A ball is thrown vertically upwards at a speed of $7.5\,\mathrm{m\,s^{-1}}$ from a height of $1.25\,\mathrm{m}$ above ground level. The ball is caught when it returns to its starting position.

   (i) Find the time it takes the ball to reach its maximum height.

   (ii) Find the maximum height of the ball above ground level.

   (iii) How long is the ball in the air for and what is its speed when it is caught?

4   A sprinter accelerates uniformly for the first 8 metres of a 100 metre race. He takes 1.5 seconds to run the first 8 metres.

   (i)   Find the acceleration in the first 1.5 seconds of the race.

   (ii) Find the speed of the sprinter after 1.5 seconds.

   (iii) The sprinter completes the 100 metres travelling at that speed. Find the total time to run the 100 metres.

   (iv) Calculate the average speed of the sprinter.

5   A particle travels with constant acceleration in a straight line. A and B are points on this line, 20 m apart. The particle is initially at A and arrives at B 40 seconds later, with a speed of 2.5 m s^{-1} moving away from A. Find the acceleration of the particle and the initial velocity of the particle, making its direction clear.

6   Two stones S_A and S_B are initially at points A and B. B is X m directly above A. S_B is dropped from rest and at the same instant S_A is projected vertically upwards with speed 25 m s^{-1}. The stones collide T seconds later and they both have the same speed V. Show that $T = 1.25$ s and find the values of V and X.

7   A car is driven with constant acceleration a m s^{-2} along a straight road. When it passes a road sign its speed is u m s^{-1}. The car travels 200 metres in the next 10 seconds after passing the sign. After 25 seconds it has a speed of 25 m s^{-1}.

    (i) Find u and a.

    (ii) What distance does the car travel in the 25 seconds after passing the sign?

8   A car is travelling at constant acceleration a m s^{-2} along a straight road. Its speed as it passes a sign is u m s^{-1}. The car travels 25 m in the 2 seconds after passing the sign and 35 m in the following 2 seconds.

    (i) Find a and u.

(ii) Find the speed of the car after 4 seconds.

(iii) How far will it travel in the next 5 seconds?

9 A particle A is projected vertically upwards, from horizontal ground, with speed $8\,\mathrm{m\,s^{-1}}$.

(i) Show that the greatest height above ground level is 3.2 m.

A second particle B is projected vertically upwards, from a point 1.4 m above the ground, with speed $u\,\mathrm{m\,s^{-1}}$. The greatest height above ground reached by B is also 3.2 m.

(ii) Find the value of u.

It is given that A and B are projected simultaneously.

(iii) Show that, at the instant when A and B are at the same height, they have the same speed and are moving in opposite directions.

10 A train, moving with constant acceleration, travels 1500 m in one minute and 2500 m in the next minute.

    (i) Find the speed of the train at the start of the first minute.

    (ii) Find the acceleration.

    (iii) Find the speed of the train at the end of the second minute.

Further practice

1 The following questions involve motion under constant acceleration.

    (i) Find a when $u = 20$, $v = 50$ and $t = 5$.
    (ii) Find a when $u = 20$, $v = 10$ and $s = 100$.

    (iii) Find s when $a = 2$, $v = 5$ and $t = 8$.
    (iv) Find v when $a = -2$, $s = 100$ and $t = 5$.

    (v) Find v when $s = 40$, $u = 10$ and $t = 5$.

2 A car travelling at $30 \, \text{m s}^{-1}$ decelerates uniformly to rest in 15 s. Find the deceleration and the distance travelled in this time.

3 A train is travelling at $40 \, \text{m s}^{-1}$ when the driver sees an obstacle across the track 100 m ahead. The brakes produce a deceleration of $10 \, \text{m s}^{-2}$. Determine whether the train hits the obstacle and, if it does, determine the speed at which it hits it.

4 A car accelerates uniformly from $10 \, \text{m s}^{-1}$ to $30 \, \text{m s}^{-1}$ over a distance of 100 m.

    (i) Find the acceleration.

    (ii) Find the speed of the car when it has travelled half the distance.

5 A ball is thrown vertically upwards and returns to its point of projection after 4 seconds. Calculate its speed of projection and the maximum height reached.

6 A pebble is thrown vertically downwards with a speed of $3.5 \, \text{m s}^{-1}$ from the top of a well, which is 22.5 m deep.

    (i) Calculate the speed of the pebble when it hits the bottom of the well.

    (ii) Find the time taken by the pebble to reach the bottom of the well.

7 A firework is projected vertically upwards at $20 \, \text{m s}^{-1}$. Find the length of time for which it is at least 5 m above the point of projection.

8 A car approaches a bend at $80 \, \text{km h}^{-1}$ and has to reduce its speed to $30 \, \text{km h}^{-1}$ in a distance of 250 m in order to go round the bend. Find the required deceleration. After going round the bend the car regains its speed in 25 seconds. Find the distance it travels in doing so.

9 The 0–60 mph time for a car is 3 seconds. Find its acceleration (assumed uniform). [Use the conversion factor 1 mile = 1610 m.]

10 The Highway Code states that for a car travelling at 30 mph (48 km h^{-1}) the stopping distance of 23 m is made up of 9 m thinking distance and 14 m braking distance.

(i) Use the thinking distance to work out the reaction time, T.

(ii) Calculate the deceleration, a, of the car as it comes to rest in 14 m.

(iii) How long does it take for the car to come to rest?

(iv) Using the values found for T and a for a car travelling at 30 mph, find the corresponding values for the thinking distance and the braking distance for a car travelling at 70 mph (112 km h^{-1}).

(v) How long does it take for a car travelling at 70 mph to come to rest?

11 A particle is projected vertically upwards from a point O at 30 m s^{-1}.

(i) Find the greatest height reached by the particle.

(ii) When the particle is at its highest point, a second particle is projected vertically upwards from O at 18 m s^{-1}. Show that the particles collide 2.5 seconds later and determine the height above O at which the collision takes place.

12 A stone is dropped from the top of a tower. One second later another stone is thrown vertically downwards with an initial speed of 12 m s^{-1}. The two stones land at the same time.

(i) Find the time taken by each stone to fall to the ground.

(ii) Find the height of the tower.

13 A particle travels with constant acceleration along a straight line. A and B are points on the line 10 m apart. The particle is initially at A and arrives at B after 40 s with a speed of 2.5 m s^{-1} moving away from A. Calculate the acceleration and initial velocity of the particle, making the directions clear.

14 A ball is thrown vertically upwards at 20 m s^{-1} from a point P. Two seconds later a second ball is also thrown vertically upwards from P with the same speed of 20 m s^{-1}.

(i) Calculate how long the first ball has been in motion when the balls meet.

(ii) Calculate the height above P at which A and B meet.

15 Two particles A and B are moving in a straight line. A starts from rest and has a constant acceleration towards B of 0.8 m s^{-2}. B starts 240 m from A at the same time and has a constant speed of 10 m s^{-1} away from A.

(i) Write down expressions for the distances travelled by A and B, t seconds after the start of the motion.

(ii) How much time does it take for A to catch up with B and how far has A travelled in this time?

16 A particle moving in a straight line with constant acceleration a passes through points O, A and B at times $t = 0$, 2 and 4 seconds respectively, where A and B are on the same side of O and OA = 10 m and OB = 50 m. Find a and the initial velocity of the particle when $t = 0$.

Past exam questions

1 A particle is projected vertically upwards with velocity 9 m s^{-1} from a point 3.15 m above horizontal ground. The particle moves freely under gravity until it hits the ground. For the particle's motion from the instant of projection until the particle hits the ground, find the total distance travelled and the total time taken. [6]

Cambridge International AS and A Level Mathematics 9709 Paper 41 Q4 June 2014

2 A particle P starts from rest at a point O on a horizontal straight line. P moves along the straight line with constant acceleration and reaches a point A on the line with speed 30 m s^{-1}. At the instant that P leaves O, a particle Q is projected vertically upwards from the point A with a speed of 20 m s^{-1}. Subsequently P and Q collide at A. Find

(i) the acceleration of P, [4]

(ii) the distance OA. [2]

Cambridge International AS and A Level Mathematics 9709 Paper 42 Q5 June 2015

3   A cyclist starts from rest at point A and moves in a straight line with acceleration $0.5\,\text{m s}^{-2}$ for a distance of $36\,\text{m}$. The cyclist then travels at constant speed for $25\,\text{s}$ before slowing down, with constant deceleration, to come to rest at point B. The distance AB is $210\,\text{m}$.

   (i) Find the total time that the cyclist takes to travel from A to B.   [5]

   $24\,\text{s}$ after the cyclist leaves point A, a car starts from rest from point A, with constant acceleration $4\,\text{m s}^{-2}$, towards B. It is given that the car overtakes the cyclist while the cyclist is moving with constant speed.

   (ii) Find the time that it takes from when the cyclist starts until the car overtakes her.   [5]

   Cambridge International AS and A Level Mathematics 9709 Paper 41 Q7 November 2015

4   A particle P is projected vertically upwards, from a point O, with a velocity of $8\,\text{m s}^{-1}$. The point A is the highest point reached by P. Find

   (i) the speed of P when it is at the mid-point of OA,   [4]

   (ii) the time taken by P to reach the mid-point of OA while moving upwards.   [2]

   Cambridge International AS and A Level Mathematics 9709 Paper 43 Q3 November 2012

5   A particle P is released from rest at the top of a smooth plane which is inclined at an angle α to the horizontal, where $\sin \alpha = \dfrac{16}{65}$. The distance travelled by P from the top to the bottom is S metres, and the speed of P at the bottom is $8\,\text{m s}^{-1}$.

   (i) Find the value of S and hence find the speed of P when it has travelled $\dfrac{1}{2}S$ metres.   [5]

   The time taken by P to travel from the top to the bottom of the plane is T seconds.

   (ii) Find the distance travelled by P at the instant when it has been moving for $\dfrac{1}{2}T$ seconds.   [2]

   Cambridge International AS and A Level Mathematics 9709 Paper 42 Q4 June 2013

▶ STRETCH AND CHALLENGE

1   A car is moving along a straight road with a speed of $10\,\text{m s}^{-1}$. It accelerates uniformly so that during the fifth second of its motion it travels a distance of $46\,\text{m}$.

   Find the acceleration and the speed of the car at the end of 5 seconds.

2   A particle A starts from the origin O with velocity $u\,\text{m s}^{-1}$ and moves along the positive x-axis with constant acceleration $a\,\text{m s}^{-2}$. Twenty seconds later, another particle B starts from O with velocity u and moves along the positive x-axis with acceleration $3a\,\text{m s}^{-2}$.

   Find the time that elapses between the start of A's motion and the instant when B has the same velocity as A, and show that A will have then travelled three times as far as B.

3   In order to determine the depth of a well, a stone is dropped into the well and the time taken for the stone to drop is measured. It is found that the sound of the stone hitting the water arrives 5 seconds after the stone is dropped. If the speed of sound is taken as $340\,\text{m s}^{-1}$, find the depth of the well.

4   A particle accelerates from rest with acceleration $3\,\text{m s}^{-2}$ to a speed $V\,\text{m s}^{-1}$. It continues at this speed for T seconds and then decelerates to rest at $1.5\,\text{m s}^{-2}$.

   The total time for the motion is one minute, and the total distance travelled is $1\,\text{km}$. Find the value of V.

5   A stone is dropped from rest from the top of a tower X m tall. During the last second it travels a distance of $0.36X$. Find the height of the tower.

3 Forces and Newton's laws of motion

3.1 Force diagrams

1 Draw force diagrams for each of the following situations. In each case, label the forces W (weight), T (tension), P (thrust), F (friction) or R (normal reaction).

(i) A ball falling to the ground (neglecting air resistance)

(ii) A ball lying on the ground

(iii) A ball being kicked on rough ground

(iv) A box at rest on a slope inclined at 30° to the horizontal

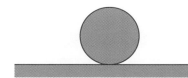

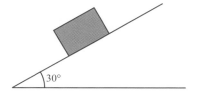

(v) A box being pulled up a slope inclined at 30° to the horizontal by a string inclined at 20° to the slope

(vi) A ladder lying on rough ground leaning on a rough wall at an angle of 30° to the wall

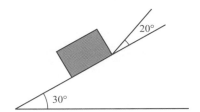

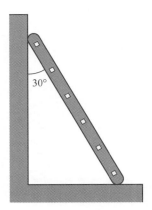

(vii) A ladder leaning against a rough wall standing on smooth ground, with a rope tied between the middle of the ladder and the wall.

(viii) A uniform plank, supported at each end by a vertical cable.

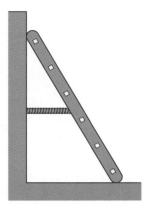

3.2 Force and motion; pulleys

1   A block of weight W_B lies on a rough horizontal table. A light inextensible string is attached to the block and runs parallel to the table, over a smooth pulley at the edge of the table and down to a weight W_A hanging freely.

   (i)   Draw the forces acting on the block and the hanging weight.

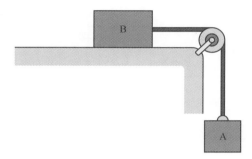

   (ii) If the block does not move, write down relations between the different forces.

   (iii) If the block is accelerating, write down the resultant force acting on the block.

2   A light smooth pulley hangs from the ceiling. A light inextensible string with two equal weights W at either end passes over the pulley.

   (i)   On the diagram show the forces acting on the pulley and the two weights.

   (ii) What is the tension in the string?

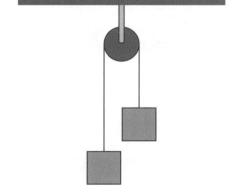

   (iii) What is the tension in the rod holding the pulley?

3 A block of mass 15 kg lies on a rough table. Two masses of 7 kg and 5 kg are attached to the block by light inextensible strings which pass over smooth pulleys at the edge of the table. The system is in equilibrium.

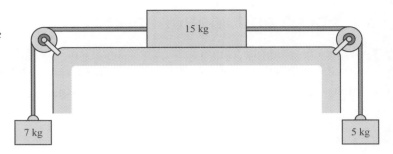

(i) Draw force diagrams for each of the three objects.

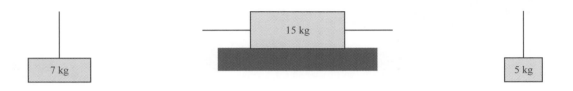

(ii) Write down the net force acting on each of these objects.

4 The figure shows a system in equilibrium. A light inextensible string passes over a fixed smooth pulley. Attached to one end of the string is a mass P of 4 kg, attached to the other end is a mass Q of 2 kg, which is itself linked through a rigid rod to a third mass R of 5 kg which is resting on the ground.

(i) Draw separate force diagrams showing all the forces acting on each of the masses P, Q, R and on the pulley.

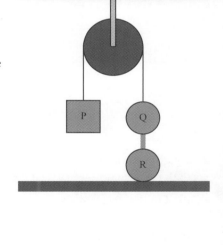

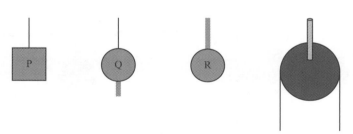

(ii) Give a reason as to why the tension in the string is the same throughout. What value is this?

(iii) Find the tension in the rod linking Q to R.

(iv) Find the normal reaction of the ground on mass R.

5 A box is at rest on the floor of a lift which is moving up and down. The normal reaction of the lift floor on the box is R and the weight of the box is W. Decide in each of the following cases whether R is greater than, less than or equal to W, and describe the net force acting on the box.

(i) The lift is moving downwards with constant velocity.

(ii) The lift is moving upwards with increasing speed.

(iii) The lift is moving upwards with decreasing speed.

(iv) The lift is moving downwards with decreasing speed.

6 Three boxes A, B and C are connected by light, rigid, horizontal rods. The boxes are lying on rough horizontal ground. A force P is applied to A in the direction BA. The frictional force on A is 25 N and that on B and C is 10 N. The boxes are moving at constant speed.

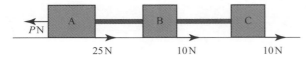

(i) Draw a diagram showing all the forces acting on the system as a whole. Hence, by considering the equilibrium of the system as a whole, find P.

The coupling between A and B exerts a force T_1 on A and the coupling between B and C exerts a force T_2 on B.

(ii) Draw diagrams showing the horizontal forces on each of the boxes.

(iii) By considering the equilibrium of A, find T_1.

(iv) By considering the equilibrium of C, find T_2.

(v) Show that the forces on B are also in equilibrium.

Further practice

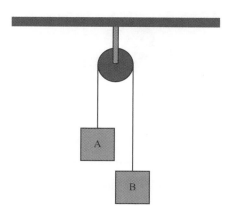

1 Two boxes A and B of masses 5 kg and 4 kg are attached at the ends of a light inextensible string passing over a smooth pulley as shown in the diagram.

(i) Draw separate diagrams showing the forces acting on A and B.

(ii) The system is released from rest. What is the net force in the direction of the motion?

2 Two particles A and B of masses 2 kg and 3 kg lie on rough horizontal ground, connected by a light rigid rod.
A force *P* is applied to A in the direction BA.

(i) Show all forces acting on A and B.

(ii) What is the net force acting on the system?

3 Three boxes of mass 1 kg, 3 kg and 5 kg are placed one on top of another with the largest at the bottom, in contact with a smooth surface. Show all the forces acting on each of the boxes.

4 The diagram shows a block of mass 10 kg connected to a light scale-pan by a light inextensible string which passes over a smooth pulley. The scale-pan holds two blocks A and B each of mass 5 kg, with B resting on top of A. The system is in equilibrium.

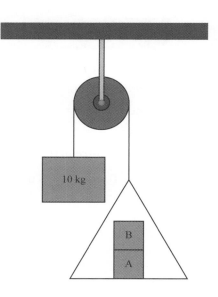

(i) On separate diagrams show all the forces acting on each of the three masses, the scale-pan and the pulley.

(ii) Find the value of the tension in the string.

(iii) Find the value of the tension in the rod holding the pulley.

(iv) Find the normal reaction between A and B.

5 Particles of masses 5 kg and 3 kg are attached to the ends of a light inextensible string. The string passes over a smooth fixed pulley. The 5 kg mass is at rest on the ground.

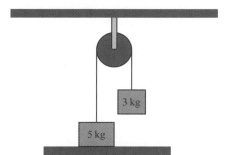

(i) Find the tension in the string.

(ii) Find the normal reaction of the floor on the 5 kg mass.

Past exam question

1

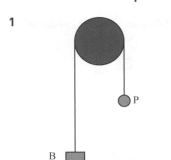

A block B of mass 5 kg is attached to one end of a light inextensible string. A particle P of mass 4 kg is attached to the other end of the string. The string passes over a smooth pulley. The system is in equilibrium with the string taut and its straight parts vertical. B is at rest on the ground (see diagram). State the tension in the string and find the force exerted on B by the ground. [3]

Cambridge International AS and A Level Mathematics 9709 paper 04 Q1 June 2009

► STRETCH AND CHALLENGE

1 A train consists of an engine of mass 50 000 kg pulling 25 trucks each of mass 10 000 kg. The force of resistance on the engine is 1500 N and that on each truck is 100 N. The train is travelling at constant speed in a straight line.

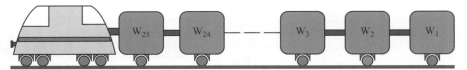

(i) Find the driving force of the engine.

(ii) Find the tension T_1 in the coupling between trucks W_1 and W_2.

(iii) Find the tension T_2 in the coupling between trucks W_2 and W_3.

(iv) Derive an expression for T_n, the force in the coupling between trucks W_n and W_{n+1}.

(v) Find the force in the coupling between the engine and truck W_{25}.

2 A smooth fixed pulley is used to lift a 25 kg box. A light inextensible string passes over the pulley and sufficient force is used to keep the box in equilibrium.

(i) Draw force diagrams showing the forces acting on the box and on the pulley.

(ii) Find the tension in the string.

(iii) Find the magnitude of the force exerted by the string on the pulley and the direction in which it is acting.

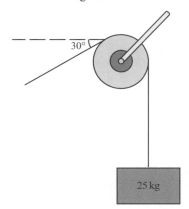

4 Applying Newton's second law along a line

4.1 Newton's second law

1 Each diagram shows the forces acting on an object and the resulting acceleration. In each case write down the equation of motion and use it to find the quantity marked with a letter.

(i)

(ii)

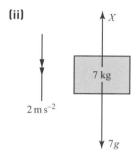

(iii)

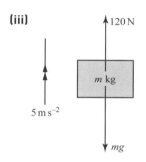

(iv)

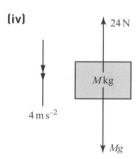

(v)

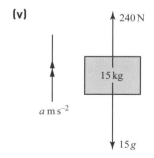

(vi)

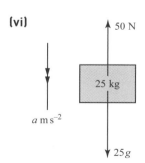

2   A box of mass 50 kg is pushed across a rough floor by a horizontal force of 100 N.
    What is the frictional force if

    (i) the box moves with constant velocity?

    (ii) the box moves with constant acceleration of $1\,\mathrm{m\,s^{-2}}$?

3   The total mass of a boy and his sailboard is 70 kg. If the wind produces a force of 200 N and the water
    a resistance of 95 N, find the acceleration of the sailboard and boy.

4   A mass of 3 kg is placed on a horizontal surface which is moving downwards with uniform acceleration.
    The normal reaction between the mass and the surface is found to be 8 N. Find the acceleration.

5   A woman of mass 65 kg is standing on the floor of a lift of mass 775 kg. The lift is descending
    with acceleration a m s^{-2}. The tension in the lift cable is 9000 N.

    (i) Calculate the value of a.

    (ii) Find the reaction of the floor on the woman.

4.2 Newton's second law applied to connected objects

1   A trailer of mass 750 kg is attached to a car of mass 1250 kg by a light rigid tow bar. The car and trailer are
    travelling along a straight road with acceleration of 1.2 m s^{-2}.

    (i) Given that the force exerted by the tow bar on the trailer is 1000 N, find the resistance
    to motion of the trailer.

    (ii) Given that the driving force of the car is 2800 N, find the resistance to motion of the car.

2   A car is towing a trailer along a straight level road. The masses of the car and the trailer are 1000 kg and 400 kg respectively. The resistance to motion of the car is 550 N and the resistance to motion of the trailer is 200 N.

At one stage of the motion, the pulling force exerted on the trailer is zero.

(i) Show that the acceleration of the trailer is $-0.5\,\mathrm{m\,s}^{-2}$.

(ii) Find the driving force exerted by the car.

(iii) Calculate the distance required to reduce the speed of the car and trailer from $22.5\,\mathrm{m\,s}^{-1}$ to $12.5\,\mathrm{m\,s}^{-1}$.

3   Two particles, A and B, are connected by a light inextensible string that passes over a smooth fixed pulley, as shown in the diagram.

The mass of A is 4 kg and the mass of B is 5 kg. The particles are released from rest in the position shown, where B is d metres above A.

(i) Find the acceleration of each particle.

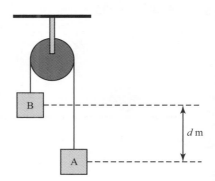

(ii) When the particles have been moving for 0.5 seconds they are at the same level. Find the speed of the particles at this time.

(iii) Find d.

4 A train consists of a locomotive and five trucks with masses and resistances to motion as shown in the figure. The engine provides a driving force of 29 000 N. All the couplings are light, rigid and horizontal.

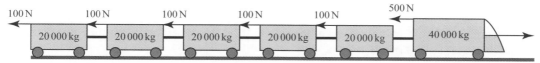

(i) Show that the acceleration of the train is $0.2\,\text{m}\,\text{s}^{-2}$.

(ii) Find the force in the coupling between the last two trucks.

The driving force is removed and the brakes are applied. This adds additional resistances of 3000 N to the locomotive and 2000 N to each truck to those shown in the figure.

(iii) Find the new acceleration of the train.

(iv) Find the force in the coupling between the last two trucks.

5 The diagram shows a block A of mass 50 kg suspended by a vertical cable. A second block B of mass 200 kg is suspended from A by means of a second vertical cable. The blocks are raised uniformly 10 metres in 10 seconds, starting from rest. Find the acceleration of the blocks and the tension in each cable.

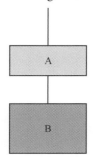

Further practice

1   A mass of 5 kg falls with acceleration $8\,\text{m}\,\text{s}^{-2}$. What resistance is acting on the mass?

2   A hot air balloon rises from the ground with uniform acceleration. It reaches a height of 200 m in 30 s. If the mass of the balloon and basket is 360 kg, find the lifting force.

3   An empty bottle of mass 2.5 kg is released from a submarine and rises with an acceleration of $0.75\,\text{m}\,\text{s}^{-2}$. The water causes a resistance of 0.5 N. Find the size of the buoyancy force.

4   A girl of mass 40 kg stands on the floor of a lift which is moving with an upward acceleration of $0.6\,\text{m}\,\text{s}^{-2}$. Calculate the magnitude of the force exerted by the floor on the girl.

5   A car of mass 1300 kg is travelling in a straight line on a horizontal road. The driving force of 3000 N acts in the direction of motion and a resistance force of 400 N opposes the motion of the car. Find the acceleration of the car.

6   An object is hung from a spring balance suspended from the roof of a lift. When the lift is descending with uniform acceleration of $2\,\text{m}\,\text{s}^{-2}$, the balance indicates a weight of 200 N. When the lift is ascending with uniform acceleration $a\,\text{m}\,\text{s}^{-2}$, the reading is 300 N. Find a.

7   A car, of mass 1400 kg, is towing a caravan of mass 800 kg along a straight horizontal road. The caravan is connected to the car by a horizontal tow bar. Resistance forces of magnitude 300 N and 550 N act on the car and caravan respectively. The acceleration of the car and caravan is $0.5\,\text{m}\,\text{s}^{-2}$.

   (i)   Find the force in the tow bar.

   (ii) Find the magnitude of the driving force.

8   Three blocks, A of mass 4 kg, B of mass 10 kg and C of mass 16 kg, are linked with rigid rods and are moving horizontally on a smooth surface. Horizontal forces of 200 N and 10 N are acting on C and A respectively.

   (i)   Draw force diagrams for each block.

   (ii) Write down separate equations of motion for each block.

   (iii) Find the acceleration and the forces in the couplings.

9   A car of mass 1200 kg is towing a caravan of mass 900 kg along a level straight road, using a rigid tow bar. The resistance to the car's motion is 200 N and the caravan experiences a resistance of 350 N.

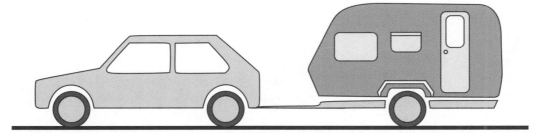

   (i)   If the driving force from the engine is 3 kN, find the tension in the tow bar and the acceleration.

   (ii) The engine is disengaged and the brakes are applied. The braking force is 500 N. Find the deceleration and the force in the tow bar, stating whether it is a tension or a compression.

10 In the following examples, two blocks with masses as shown in the diagrams are connected by a light inextensible string which passes over a smooth fixed pulley. The blocks are released from rest with the heavier of the two masses a distance above ground level, as given in the diagrams.

(i)

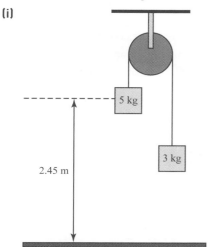

(ii)

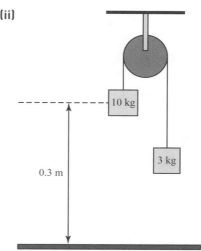

(a) Find the acceleration and the tension in the string in each case.

(b) Find, in each case, the speed of the masses when the heavier mass hits the ground.

(c) Find, in each case, how far the lighter mass rises after the heavier mass hits the floor and the string becomes slack.

11 Two boxes are descending vertically, supported by a parachute. Box A has a mass of 60 kg. Box B has a mass of 35 kg and hangs from box A by a light rigid wire. At one stage the boxes are decelerating, with the parachute exerting an upward force of 1050 N on box A. The acceleration of the boxes is $a\,\text{m}\,\text{s}^{-2}$ upwards and the tension in the wire is T N.

(i) Draw separate force diagrams showing all the forces acting on box A and box B.

(ii) Write down separate equations of motion for box A and box B.

(iii) Calculate a and T.

12 A particle P of mass 3 kg is lying on a smooth horizontal table top which is 1.5 m above the floor. A light inextensible string of length 1 m connects P to a particle Q also of mass 3 kg which hangs freely over a smooth pulley at the edge of the table. Initially P is held 0.5 m from the pulley when the system is released from rest.

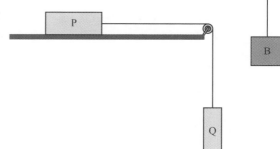

(i) Find the speed of P when it reaches the pulley.

(ii) Find the tension in the string.

13 Block A of mass 25 kg lies on a smooth table. Two blocks, B of mass 12 kg and C of mass 5 kg, are attached to block A by light inextensible strings which pass over smooth pulleys at the edges of the table.

(i) Draw separate force diagrams for A, B and C.

(ii) Write down the equation of motion for each block.

(iii) Find the values of the acceleration and the tensions in the strings.

(iv) The table is not smooth and in fact the acceleration is half the value found in (iii). Find the frictional force that would produce this result.

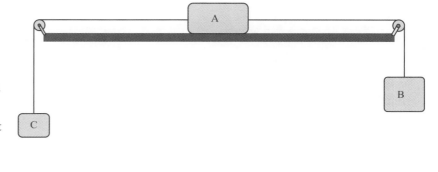

14 A train has a mass of 250 tonnes. The engine exerts a pull of 50 kN. The resistance to motion is 1% of the weight of the train. The braking force of the engine is 20% of the weight. The train starts from rest and accelerates uniformly until it reaches a speed of $10\,\text{m}\,\text{s}^{-1}$. At this point the brakes are applied until the train stops. Find the time taken for the train to stop once the brakes are applied, to the nearest second.

15 Particles A and B are attached to the ends of a light inextensible string. The string passes over a smooth pulley. The particles are released from rest, with the string taut, and A and B at the same height above a horizontal floor. In the subsequent motion A descends with acceleration $2\,\text{m}\,\text{s}^{-2}$ and strikes the floor 0.5 s after being released. It is given that B never reaches the pulley.

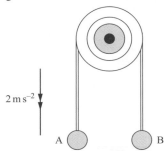

$2\,\text{m}\,\text{s}^{-2}$

A   B

(i) Calculate the distance A moves before reaching the ground and the speed of A immediately before it hits the floor.

(ii) Show that B rises a further 0.05 m after A strikes the floor and calculate the total length of time during which B is rising.

(iii) Before A strikes the floor the tension in the string is 6 N. Calculate the mass of A and the mass of B.

(iv) The pulley has mass 0.5 kg and is held in a fixed position by a light vertical rod. Calculate the tension in the rod

(a) immediately before A strikes the floor

(b) immediately after A strikes the floor.

Past exam questions

1 Two particles A and B have masses 0.12 kg and 0.38 kg respectively. The particles are attached to the ends of a light inextensible string which passes over a fixed smooth pulley. A is held at rest with the string taut and both straight parts of the string vertical. A and B are each at a height of 0.65 m above horizontal ground (see diagram). A is released and B moves downwards.

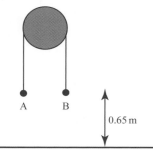

A   B

0.65 m

Find

(i) the acceleration of B while it is moving downwards, [2]

(ii) the speed with which B reaches the ground and the time taken for it to reach the ground. [3]

B remains on the ground while A continues to move with the string slack, without reaching the pulley. The string remains slack until A is at a height of 1.3 m above the ground for a second time. At this instant A has been in motion for a total of T s.

(iii) Find the value of T and sketch the velocity–time graph for A for the first T s of its motion. [3]

(iv) Find the total distance travelled by A in the first T s of its motion. [2]

Cambridge International AS and A Level Mathematics 9709 Paper 43 Q7 June 2012

2 Particles A and B have masses 0.32 kg and 0.48 kg respectively. The particles are attached to the ends of a light inextensible string which passes over a small smooth pulley fixed at the edge of a smooth horizontal table. Particle B is held at rest on the table at a distance of 1.4 m from the pulley. A hangs vertically below the pulley at a height of 0.98 m above the floor (see diagram). A, B, the string and the pulley are all in the same vertical plane. B is released and A moves downwards.

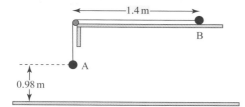

(i) Find the acceleration of A and the tension in the string. [5]

A hits the floor and B continues to move towards the pulley. Find the time taken, from the instant that B is released, for

(ii) A to reach the floor, [2]

(iii) B to reach the pulley. [3]

Cambridge International AS and A Level Mathematics 9709 Paper 43 Q7 November 2012

3 Particles P and Q are attached to opposite ends of a light inextensible string which passes over a smooth fixed pulley. The system is released from rest with the string taut, with its straight parts vertical, and with both particles at a height of 2 m above horizontal ground. P moves vertically downwards and does not rebound when it hits the ground. At the instant that P hits the ground, Q is at the point X, from where it continues to move vertically upwards without reaching the pulley. Given that P has mass 0.9 kg and that the tension in the string is 7.2 N while P is moving, find the total distance travelled by Q from the instant it first reaches X until it returns to X. [6]

Cambridge International AS and A Level Mathematics 9709 Paper 43 Q3 November 2011

4

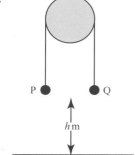

Figure 1

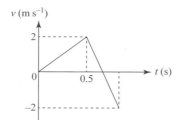

Figure 2

Two particles P and Q have masses m kg and $(1 - m)$ kg respectively. The particles are attached to the ends of a light inextensible string which passes over a smooth fixed pulley. P is held at rest with the strings taut and both straight parts of the string vertical. P and Q are each at a height of h m above horizontal ground (see Figure 1). P is released and Q moves downwards. Subsequently Q hits the ground and comes to rest. Figure 2 shows the velocity–time graph for P while Q is moving downwards or is at rest on the ground.

(i) Find the value of h. [2]

(ii) Find the value of m, and find also the tension in the string while Q is moving. [6]

(iii) The string is slack while Q is at rest on the ground. Find the total time from the instant that P is released until the string becomes taut again. [3]

Cambridge International AS and A Level Mathematics 9709 Paper 42 Q6 June 2015

▶ STRETCH AND CHALLENGE

1 The engine of a goods train has mass 50 tonnes and is pulling a convoy of 25 trucks, each of mass 8 tonnes. The resistive forces are 2500 N on the engine and 250 N on each truck. The driving force of the engine is 40 kN and the train is travelling along a level straight track.

(i) Find the acceleration of the train.

(ii) Find the force in the coupling between the engine and the first truck.

(iii) Show that the force T_n in the coupling between the nth truck and the $(n + 1)$th truck is given by
$T_n = 31\,250 - 1250n$.

(iv) Hence, or otherwise, find the force in the coupling between the last two trucks.

2 A light inextensible string passes over a smooth pulley and carries at one end a particle A of mass 5 kg and at the other end a light smooth pulley over which passes a second light inextensible string carrying particles B of mass 2 kg and C of mass 3 kg at its ends.

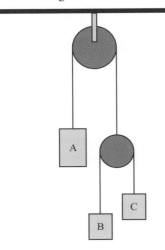

(i) Find the acceleration of A when the particles move vertically under gravity.

(ii) Find by how much the mass of A must be reduced in order that A can remain at rest while the other two particles are in motion.

3 Block A of mass 2 kg is connected to a light scale-pan by a light inextensible string which passes over a smooth fixed pulley. The scale-pan holds two blocks, B and C, of masses 0.5 kg and 1 kg, as shown in the figure.

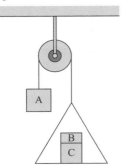

(i) Draw diagrams showing all the forces acting on each of the three blocks and on the scale-pan.

(ii) Write down equations of motion for each of A, B, C and the scale-pan.

(iii) Find the acceleration of the system, the tension in the string, the normal force between B and C and the normal force between C and the scale-pan.

5 Vectors

1 Find the magnitude and direction of each of the following vectors, giving the direction from the positive x-axis. Vectors **i** and **j** are unit vectors along the x and y-directions respectively.

(i) $5\mathbf{i} + 12\mathbf{j}$

(ii) $\mathbf{i} - 2\mathbf{j}$

(iii) $4\mathbf{i} - 3\mathbf{j}$

(iv) $-2\mathbf{i} + 5\mathbf{j}$

2 Vectors **i** and **j** are unit vectors in the east and north directions respectively. Find the magnitude and direction of the following vectors, giving the direction as a bearing.

(i) $\mathbf{i} + \mathbf{j}$

(ii) $3\mathbf{i} - 5\mathbf{j}$

(iii) $20\mathbf{i} - 4.5\mathbf{j}$

(iv) $-30\mathbf{i} + 5.5\mathbf{j}$

3   Three forces of magnitudes 5 N, 6 N and 8 N are acting at the point O as shown in the diagram. Find the magnitude and direction of the resultant of the three forces.

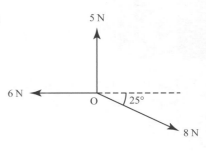

4   Two forces $\mathbf{F_1} = \begin{pmatrix} 5 \\ -8 \end{pmatrix}$ and $\mathbf{F_2} = \begin{pmatrix} 3 \\ 2 \end{pmatrix}$ are acting on a particle of mass 2 kg.

(i)   Find the resultant force acting on the particle and hence the acceleration of the particle.

(ii)   When $t = 10$ seconds, the velocity of the particle is given by $\mathbf{v} = \begin{pmatrix} 30 \\ -10 \end{pmatrix}$. Find the initial velocity $\mathbf{u}$.

(iii)   Find the time when the speed of the particle is $10\,\mathrm{m\,s^{-1}}$.

5   A particle is moving with constant acceleration $\mathbf{a} = \begin{pmatrix} 3 \\ -5 \end{pmatrix}$. It passes point O at $t = 0$ with initial velocity $\mathbf{u} = \begin{pmatrix} -1 \\ 3 \end{pmatrix}$. The unit vectors $\begin{pmatrix} 1 \\ 0 \end{pmatrix}$ and $\begin{pmatrix} 0 \\ 1 \end{pmatrix}$ are pointing due east and due north respectively.

(i)   Show that after 1 second the particle is moving in the direction SE and find its speed.

(ii)   Calculate the bearing of the particle from O after it has been moving for 2.5 seconds.

6 Three forces $\mathbf{F}_1$ of magnitude 250 N at an angle of 35° to the positive x-direction, $\mathbf{F}_2$ of magnitude 350 N along the positive x-direction and $\mathbf{F}_3$ of magnitude 200 N at an angle of 20° below the positive x-direction, are applied to a packing case in order to move it.

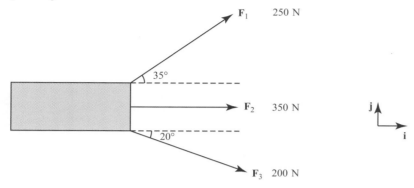

(i) Express each of the three forces in terms of their components.

(ii) Find the magnitude and direction of the resultant force.

(iii) A fourth force $\mathbf{F}_4 = 120\mathbf{i} + P\mathbf{j}$ is now applied and the case now moves at constant speed along the positive x-direction. Find P.

7 Vector $\mathbf{i}$ is a unit vector pointing due east and $\mathbf{j}$ is a unit vector pointing due north.

(i) If $\mathbf{F} = 5\mathbf{i} - 8\mathbf{j}$, find the magnitude of $\mathbf{F}$ and its direction expressed as a bearing.

(ii) If $\mathbf{G} = 7.5\mathbf{i} - 12\mathbf{j}$, express $\mathbf{G}$ in terms of $\mathbf{F}$.

(iii) If $\mathbf{F}_1 = 3\mathbf{i} - 5\mathbf{j}$ and $\mathbf{F}_2 = 7\mathbf{i} + Q\mathbf{j}$, find Q such that $\mathbf{F}_1 + \mathbf{F}_2$ is parallel to $\mathbf{F}$.

Further practice

1 The coordinates of A and B are (3, 2) and (–2, 5). Express $\overrightarrow{AB}$ in terms of its magnitude and direction.

2 Find the coordinates of A if $\overrightarrow{AB} = \begin{pmatrix} 2 \\ 5 \end{pmatrix}$ and B is the point (7, 1).

3 If $\mathbf{p} = \begin{pmatrix} 2 \\ -1 \end{pmatrix}$ and $\mathbf{q} = \begin{pmatrix} 3 \\ 5 \end{pmatrix}$ find $|\mathbf{p} + \mathbf{q}|$ and $|\mathbf{p} - \mathbf{q}|$.

4 Vectors $\mathbf{a}$ and $\mathbf{b}$ are perpendicular with $|\mathbf{a}| = 5$ and $|\mathbf{b}| = 12$. Find $|\mathbf{a} + \mathbf{b}|$ and $|\mathbf{a} - \mathbf{b}|$.

5 Vectors $\mathbf{a}$ and $\mathbf{b}$ are such that $|\mathbf{a}| = 13$, $|\mathbf{b}| = 19$ and $|\mathbf{a} + \mathbf{b}| = 24$. Find $|\mathbf{a} - \mathbf{b}|$.

6 The three forces $\mathbf{F}_1 = \begin{pmatrix} -7 \\ 4 \end{pmatrix}$, $\mathbf{F}_2 = \begin{pmatrix} 10 \\ -9 \end{pmatrix}$ and $\mathbf{F}_3$ are in equilibrium. Find the magnitude and direction of $\mathbf{F}_3$.

7 Find the magnitude and direction of the resultant of these vectors.

   (i) $\mathbf{p} = 3\mathbf{i} + 2\mathbf{j}$ and $\mathbf{q} = 5\mathbf{i} - 7\mathbf{j}$, where $\mathbf{i}$ is a unit vector in the x-direction and $\mathbf{j}$ is a unit vector in the y-direction.

   (ii) $\mathbf{a} = \begin{pmatrix} -2 \\ 1 \end{pmatrix}$, $\mathbf{b} = \begin{pmatrix} 3 \\ -7 \end{pmatrix}$ and $\mathbf{c} = \begin{pmatrix} 9 \\ 2 \end{pmatrix}$

8 Two forces of magnitude 5 N and 4 N act at a point. The angle between the two forces is 60°.

   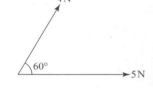

   (i) Calculate the magnitude of the resultant force.

   (ii) Calculate the angle between the resultant and the 5 N force.

9 Find the magnitude and direction of the resultant of two forces of magnitude 26 N and $5\sqrt{5}$ N acting in the directions $12\mathbf{i} - 5\mathbf{j}$ and $2\mathbf{i} + \mathbf{j}$ respectively.

10 A force $\mathbf{P}$ is given in component form by $\mathbf{P} = \begin{pmatrix} 36 \\ -4.25 \end{pmatrix}$.

   (i) Find the magnitude and direction of $\mathbf{P}$.

   A second force $\mathbf{Q}$ of magnitude 87 N acts in the same plane at 90° anticlockwise from $\mathbf{P}$. The resultant of these two forces has a magnitude R and makes an angle θ with the positive x-axis.

   (ii) Find R and θ.

Past exam questions

1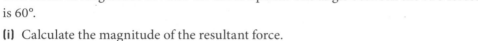

Four coplanar forces act at a point. The magnitudes of the forces are 5 N, 4 N, 3 N and 7 N and the directions in which the forces act are shown in the diagram. Find the magnitude and direction of the resultant of the four forces.

[6]

Cambridge International AS and A Level Mathematics 9709 Paper 41 Q3 June 2014

2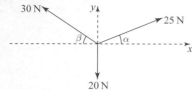

Three coplanar forces act at a point. The magnitudes of the forces are 20 N, 25 N and 30 N, and the directions in which the forces act are as shown in the diagram, where $\sin \alpha = 0.28$ and $\cos \alpha = 0.96$, and $\sin \beta = 0.6$ and $\cos \beta = 0.8$.

(i) Show that the resultant of the three forces has a zero component in the x-direction. [2]

(ii) Find the magnitude and direction of the resultant of the three forces. [2]

(iii) The force of magnitude 20 N is replaced by another force. The effect is that the resultant force is unchanged in magnitude but reversed in direction. State the magnitude and direction of the replacement force. [1]

Cambridge International AS and A Level Mathematics 9709 Paper 42 Q2 November 2014

▶ STRETCH AND CHALLENGE

1   The magnitude of the resultant of two forces **P** and **Q** is equal to the magnitude of **P**. The magnitude of the resultant of 2**P** and **Q** is equal to $\sqrt{3}$ times the magnitude of **P**.

Find the magnitude of **Q** and show that **Q** makes an angle of 120° with **P**.

2   Two forces of magnitude 5 N and 8 N act at an angle θ such that $\sin \theta = 0.25$. Find the magnitude of the two possible resultants.

3   The vertices of a triangle ABC have position vectors **a**, **b** and **c**. Points M, N and P are the midpoints of BC, CA and AB respectively, and G is the point which is $\frac{2}{3}$ of the way along AM.

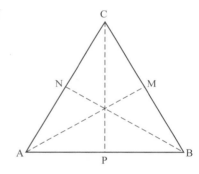

   (i) Find the following in terms of **a**, **b** and **c**.

   (a) $\overrightarrow{AM}$

   (b) $\overrightarrow{AG}$

   (c) $\overrightarrow{BN}$

   (d) $\overrightarrow{CP}$

   (e) the position vector of G.

   (ii) Show that G is also $\frac{2}{3}$ of the way along BN and CP.

4   Three tugs are pulling a ship due north into a harbour. The ropes attaching the ship to the tugs are in the direction NW, N15°W and N30°E. The tensions in the first two ropes are 200 kN and 1000 kN.

   Find the tension in the third rope and the resultant pull on the ship.

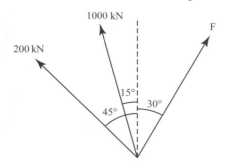

6 Forces in equilibrium and resultant forces

6.1 Finding resultant forces

1 For each diagram, find

(a) the resultant force in terms of its components

(b) the magnitude and direction of the resultant.

(i)

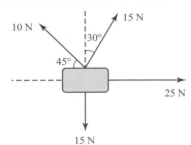

(ii)

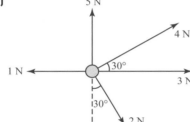

(iii)

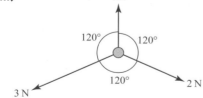

6.2 Forces in equilibrium

1 In each of the force diagrams, a particle is kept in equilibrium under the action of the forces shown. In each case find the values of the unknown forces and angles marked by letters.

(i)

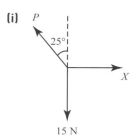

(ii)

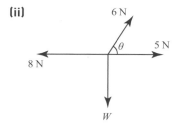

(iii)

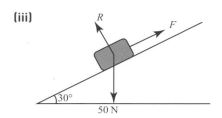

(iv)

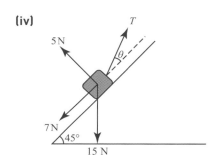

(v)

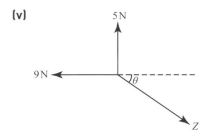

(vi)

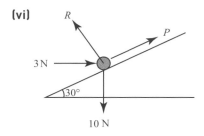

2 A particle of mass 7 kg is suspended by two strings of length 0.25 m and 0.6 m to two points A and B which are on the same level and 0.65 m apart.

Find the tensions in the two strings.

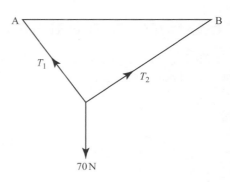

3 A particle of mass 3 kg rests on a smooth plane inclined at an angle θ to the horizontal. It is attached to a light inextensible string which passes over a smooth pulley at the top of the plane. At the other end of the string is a mass of 2 kg which is hanging freely. If the system is in equilibrium, find the angle θ, the tension in the string and the normal reaction between the 3 kg particle and the plane.

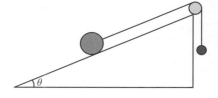

4 A particle is suspended by two light inextensible strings and hangs in equilibrium. One string is inclined at 35° to the horizontal and has a tension of 40 N. The second string is inclined at 55° to the horizontal.

(i) Find the tension in the second string.

(ii) Find the mass of the particle.

6.3 Newton's second law in two dimensions

1   A block of mass 12 kg is placed on a smooth plane inclined at 40° to the
    horizontal. It is connected by a light inextensible string, which passes over a
    smooth pulley at the top of the plane, to a mass of 7 kg hanging freely. Find the
    common acceleration and the tension in the string.

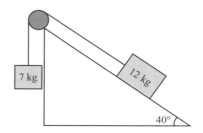

2   A car is towing a trailer along a straight road down a slope inclined at 5° to the horizontal. The masses of the car
    and trailer are 950 kg and 350 kg respectively. The resistance to motion of the car is 500 N and that for the trailer
    is 200 N. The driving force of the car is 1000 N.

    (i) Find the acceleration of the car.

    (ii) Find the pulling force exerted on the trailer.

3   A slide consists of a sloping section followed by a horizontal section. Both sections are 5 m long. The sloping
    section is inclined at an angle θ to the horizontal, with $\sin \theta = 0.28$. A child of mass 25 kg starts from rest at the
    top of the slide and is subject to a constant resistance of 35 N along both sections of the slide.

    (i) Calculate the child's acceleration down the slope.

    (ii) Calculate the speed of the child at the bottom of the slope.

(iii) Calculate the child's deceleration on the horizontal section.

(iv) Calculate the speed of the child at the end of the horizontal section.

4   A locomotive of mass 35 tonnes is pulling two trucks each of mass 15 tonnes along a horizontal straight track. The locomotive is subject to a resistance of 1000 N and each truck to a resistance of 400 N.

(i)   Find the driving force of the locomotive if the train is travelling at constant speed.

(ii) Find the force in each of the couplings.

The train now comes to an incline of 5° to the horizontal and starts to ascend.

(iii) If the locomotive maintains the same driving force, calculate the deceleration of the train and the new forces in the couplings.

5 Two boxes, A of mass 4 kg and B of mass 2.5 kg, are linked by a light rigid rod. A force of 50 N is pulling the boxes up a slope inclined at an angle θ to the horizontal such that $\sin\theta = 0.1$.

Resistances to the motion of the boxes are 10.5 N for A and 7.5 N for B.

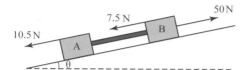

Find the acceleration of the boxes and the force in the coupling, stating whether it is a tension or a thrust.

6 A car of mass 900 kg is towing a trailer of mass 350 kg along a slope of 1 in 100 (i.e. at an angle θ to the horizontal and $\sin\theta = \dfrac{1}{100}$). The driving force of the engine is 2000 N and there are resistances to the motion of 500 N on the car and 200 N on the trailer.

Find the acceleration and the tension in the tow bar.

Further practice

1 A particle of mass 6 kg is held in equilibrium on a smooth inclined plane at 35° to the horizontal by an inextensible string which makes an angle of 15° with the slope.

(i) Find the tension in the string.

(ii) Find the normal reaction of the plane on the particle.

2 A force of 0.5 N making an angle of 20° above the horizontal is applied to a particle of mass 25 grams hanging at the end of a string. If the mass is in equilibrium, find the tension in the string and the angle which the string makes with the vertical.

3 A particle of mass 7 kg is attached at the end of a string. The other end of the string is fixed. An upward vertical force of 5 N and a horizontal force of 21 N act upon the particle, so that it rests in equilibrium with the string at an angle θ to the vertical. Calculate the tension in the string and the angle θ.

4 A box of mass 25 kg is supported by two light strings AC and BC that are tied to the box at C. AC and BC make angles of 45° and 30° with the horizontal. There is also a horizontal force of 20 N acting at C as shown in the diagram. Find the tensions in the two strings.

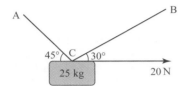

5 A box of mass 30 kg is on a smooth horizontal surface, as shown in the figure. A light string AB is attached to the surface at A and to the box at B. AB makes an angle of 40° to the vertical. Another light string is attached to the box at C; this string is inclined at 25° to the horizontal and the tension in it is 50 N. The box is in equilibrium.

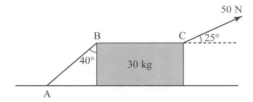

(i) Calculate the tension in the string AB.

(ii) Calculate the normal reaction of the floor on the box.

The string at C is replaced by another string which is inclined at 25° below the horizontal with the same tension of 50 N.

(iii) Explain why this has no effect on the tension in the string AB.

(iv) How much larger is the normal reaction?

6 The diagram shows a sign attached to a point A. It is supported by two rigid light rods AB and AC. AB is horizontal and AC makes an angle θ with the horizontal where $\sin \theta = 0.75$. The mass of the sign is 25 kg. Find the forces in rods AB and AC, stating whether they are in tension or compression.

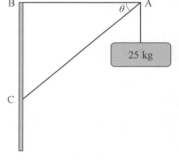

7 A sign of mass 5 kg is hung from the ceiling of a shop by two light strings, each making an angle of 25° with the vertical as shown in the diagram.

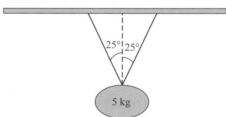

(i) Show that the tension is the same in each string.

(ii) Find the tension in each string.

(iii) If the tension in each string exceeds 75 N, the string will break. Find the mass of the heaviest sign which can be hung in this way.

8   A particle is in equilibrium under the set of forces shown in the diagram.

Show that $P = \left(3\sqrt{2} - 2\right)\dfrac{W}{2}$

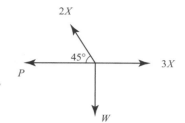

9   A uniform rod AB of weight W rests in equilibrium with the end A in contact with a smooth vertical wall and the end B in contact with a smooth plane inclined at 45° to the wall. Find the reactions at A and B in terms of W.

10 Two boxes of masses 15 kg and 12 kg are held by light strings AB, BC and CD. As shown in the figure, AB makes an angle α with the horizontal and is fixed at A. Angle α is such that $\sin\alpha = 0.6$ and $\cos\alpha = 0.8$. BC is horizontal and CD makes an angle β with the horizontal.

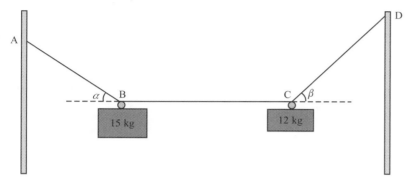

(i)   By considering the equilibrium of B, find the tension in string AB and show that the tension in BC is equal to 200 N.

(ii) By considering the equilibrium of point C, find β and the tension in CD.

11 A block of mass 75 kg is in equilibrium on smooth horizontal ground with one end of a light string attached to its upper edge. The string passes over a smooth pulley, with a block of mass m kg attached at the other end.

The part of the string between the pulley and the block makes an angle of 65° with the horizontal. A horizontal force F is also acting on the block.

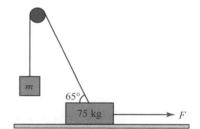

(i) T is the tension in the string and R is the normal reaction of the floor on the block. Find a relationship between T and R.

It is given that the block is on the point of lifting off the ground.

(ii) Find T and m.

(iii) Find F.

12 Two particles A and B rest on the inclined faces of a fixed triangular wedge as shown in the diagram. A and B are connected by a light inextensible string which passes over a light smooth pulley at C. The faces of the wedge are smooth. A and B have the same mass, 5 kg.

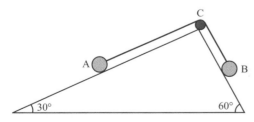

Find the acceleration of the system, the tension in the string and the force exerted by the string on the pulley at C.

13 A car of mass 1050 kg pulls a trailer of mass 450 kg up a road inclined at an angle θ to the horizontal, where $\sin\theta = 0.125$. The resistance to motion for both car and trailer is 0.25 N per kg.

   (i) Find the driving force exerted by the engine and the tension in the coupling if the car and trailer are travelling at constant speed.

   (ii) Find the driving force exerted by the engine and the tension in the coupling if the car and trailer are accelerating at $0.8\,\mathrm{m\,s}^{-2}$.

14 A block of mass 5 kg is being accelerated at a rate of $1\,\mathrm{m\,s}^{-2}$ up a smooth plane inclined at an angle θ to the horizontal, such that $\sin\theta = 0.26$. A light inelastic string is attached to the block, passes over a smooth pulley and supports a mass of m kg which is hanging freely.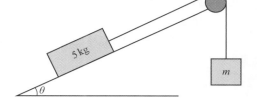

   (i) Find m and the tension in the string.

   (ii) If $m = 3.5$ kg, find the acceleration and the tension in the string in this case.

15 A large box of mass 300 kg is being pulled by three forces as shown in the diagram. The angle θ is chosen so that the resultant of the three forces acts along the **i**-direction.

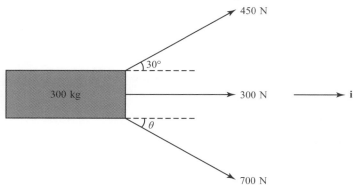

   (i) Find θ and the resultant of the three forces.

   With this resultant force, the box moves with constant acceleration and travels 1 m from rest in 5 s.

   (ii) Find the magnitude of the frictional force.

   When the speed of the box is $0.5\,\mathrm{m\,s}^{-1}$ it comes to a point on the floor where the friction is 250 N smaller. The pulling forces are the same.

   (iii) Find the velocity of the box when it has moved a further 1.5 m.

16 (i) Find the resultant of the set of six forces whose magnitudes and directions are shown in the figure.

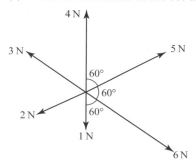

   The forces are acting on a particle P of mass 5 kg which is initially at rest at O.

   (ii) How fast is P moving after 3 s and how far from O is it now?

17 A box of mass 50 kg is sliding down a slope inclined at 20° to the horizontal. The box starts from rest and reaches a speed of 4.5 m s^{-1} after 2 seconds.

Calculate the frictional force between the box and the slope.

18 Two toy trucks are travelling down a slope inclined at an angle of 5° to the horizontal. Truck A has mass 0.5 kg, truck B has mass 0.35 kg. The trucks are linked by a light rigid rod which is parallel to the slope. The resistances to motion of the trucks are 0.25 N for truck A and 0.15 N for truck B. The initial speed of the trucks is 2 m s^{-1}.

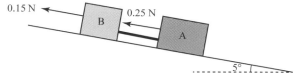

(i) Calculate the speed of the trucks after 3 seconds.

(ii) Calculate the force in the rod connecting the two trucks, stating whether the rod is in tension or in thrust.

19 A train of mass 200 tonnes is travelling uniformly on level ground at 10 m s^{-1} when it begins an ascent of 1 in 50. The driving force exerted by the engine is equal to 25 kN and the resistance force on the train is a constant 10 kN. Find how far the train climbs before it comes to a standstill.

Past exam questions

1 A smooth ring R of mass 0.16 kg is threaded on a light inextensible string. The ends of the string are attached to fixed points A and B. A horizontal force of magnitude 11.2 N acts on R, in the same vertical plane as A and B. The ring is in equilibrium. The string is taut with angle ARB = 90°, and the part AR of the string makes an angle of $\theta°$ with the horizontal (see diagram). The tension in the string is T N.

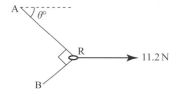

(i) Find two simultaneous equations involving $T \sin \theta$ and $T \cos \theta$. [3]

(ii) Hence find T and θ. [3]

Cambridge International AS and A Level Mathematics 9709 Paper 43 Q2 June 2012

2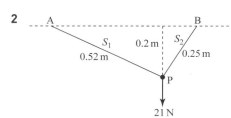

A particle P of weight 21 N is attached to one end of each of two light inextensible strings, S$_1$ and S$_2$, of lengths 0.52 m and 0.25 m respectively. The other end of S$_1$ is attached to a fixed point A, and the other end of S$_2$ is attached to a fixed point B at the same horizontal level as A. The particle P hangs in equilibrium at a point 0.2 m below the level of AB with both strings taut (see diagram). Find the tension in S$_1$ and the tension in S$_2$. [6]

Cambridge International AS and A Level Mathematics 9709 Paper 43 Q4 November 2012

3 Coplanar forces of magnitude 58 N, 31 N and 26 N act at a point in the directions shown in the diagram.

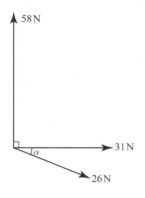

Given that $\tan \alpha = \frac{5}{12}$, find the magnitude and direction of the resultant of the three forces. [6]

Cambridge International AS and A Level Mathematics 9709 Paper 43 Q2 November 2011

4

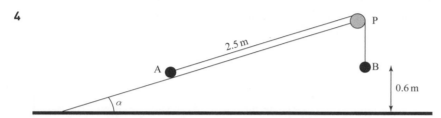

Particles A of mass 0.26 kg and B of mass 0.52 kg are attached to the ends of a light inextensible string. The string passes over a small smooth pulley P which is fixed at the top of a smooth plane. The plane is inclined at an angle α to the horizontal, where $\sin \alpha = \frac{16}{65}$ and $\cos \alpha = \frac{63}{65}$. A is held at rest at a point 2.5 metres from P, with the part AP of the string parallel to a line of greatest slope of the plane. B hangs freely below P at a point 0.6 m above the floor (see diagram). A is released and the particles start to move. Find

(i) the magnitude of the acceleration of the particles and the tension in the string, [5]

(ii) the speed with which B reaches the floor and the distance of A from P when A comes to instantaneous rest. [6]

Cambridge International AS and A Level Mathematics 9709 Paper 42 Q7 June 2013

▶ STRETCH AND CHALLENGE
· ·

1 A particle of mass m rests on a smooth plane inclined at 30° to the horizontal. A force P, inclined at an angle θ to the plane, holds the particle in equilibrium.

If the normal reaction is given by $R = \dfrac{mg}{\sqrt{3}}$, find the value of θ and show that $P = \dfrac{mg}{\sqrt{3}}$

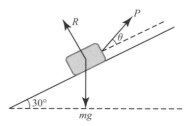

2 A sleigh is found to travel with uniform speed down a slope of 1 in 50. If the sleigh starts from the bottom of the slope with a speed of $2\,\mathrm{m\,s}^{-1}$ how far will it travel up the slope before coming to rest?

3 Three light inextensible strings have a particle attached to one of their ends. The other ends are tied together at A. The strings are in equilibrium with two of them passing over smooth pulleys and the particles are hanging freely. The weight of the particles and the angles between the sloping parts and the vertical are shown in the figure.

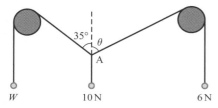

Find all possible values of W and θ.

7 General motion in a straight line

7.1 Using differentiation

1 A particle moves along the x-axis. Its position at time t is given by

$$x = 21 - 9t + 6t^2 - t^3$$

   (i) Find an expression for the velocity at time t.

   (ii) Show that for the first second the particle moves in the negative direction, and for the next two seconds it moves in the positive direction.

   (iii) Find also the acceleration at the instants when the particle is at rest.

2 The position s of a moving point A at time t seconds is given by

$$s = t^4 - t^2 + 2$$

   (i) Find an expression for the velocity.

   (ii) Use the equations to write down the initial position and velocity.

   (iii) Find the times and positions when the velocity is zero.

3 The displacement x m of a particle from the origin O is given by

$$x = 15 + 12t + 3t^2 - 2t^3; \quad -3 \leqslant t \leqslant 5.$$

(i) Write down the displacement of the particle when $t = 0$.

(ii) Find an expression in terms of t for the velocity v m s^{-1}.

(iii) Find an expression in terms of t for the acceleration a m s^{-2}.

(iv) Find the maximum value of v in the interval $-3 \leqslant t \leqslant 5$.

(v) Determine the number of times the particle passes through the origin in the interval $-3 \leqslant t \leqslant 5$.

(vi) Find the total distance travelled in the interval $-3 \leqslant t \leqslant 5$.

7.2 Finding displacement from velocity and finding velocity from acceleration

1 Find the displacement s m of a particle at time t s if the velocity v m s^{-1} is given by

(i) $v = 5t - 2$ and $s = 3$ when $t = 0$

(ii) $v = 3t^2 + 4t$ and $s = 2$ when $t = 1$

(iii) $v = 1 - 3t$ and $s = -2$ when $t = 0$.

2 Find the velocity $v\,\mathrm{m\,s^{-1}}$ and displacement $s\,\mathrm{m}$ of a particle at time $t\,\mathrm{s}$, if its acceleration $a\,\mathrm{m\,s^{-2}}$ is given by

   (i) $a = 12t - 8$ and $s = 5$ and $v = 3$ when $t = 0$

   (ii) $a = 3.5t^{1.5} - 3$ and $s = 3$ and $v = -2$ when $t = 1$

   (iii) $a = 0.6t^2 - 0.3t + 1$ and $s = 0.5$ and $v = 1$ when $t = 1$.

3 A particle P is moving in a straight line. At time t seconds after starting from O, its velocity v is given by

   $v = t^2(3 - t)$

   (i)   Find the values of t when the acceleration is zero.

   (ii)   At what time does the particle come to instantaneous rest?

   At this time the particle has reached its furthest point A from O. It then reverses direction and travels back towards O.

   (iii) Find the distance OA.

   (iv) Find the time taken for the particle to return to O.

4 A sprinter starts from rest at time $t = 0$, and runs in a straight line. For $0 \leqslant t \leqslant 3$ the sprinter has a velocity given by $v = 2.7t^2 - 0.6t^3$. For $3 \leqslant t \leqslant 23$, the sprinter runs at a constant speed of $8.1\,\mathrm{m\,s^{-1}}$. For $t > 23$ the sprinter decelerates at a constant rate of $0.2\,\mathrm{m\,s^{-2}}$.

(i) Find the distance travelled by the sprinter in the first 3 seconds.

(ii) Find the time the sprinter takes to run 100 metres.

(iii) Find the time the sprinter takes to run 200 metres.

5 The acceleration, $a\,\mathrm{m\,s^{-2}}$, of a particle t s after starting from rest is given by $a = 3t - 2$.

(i) Show that the particle returns to its starting point after $2\,\mathrm{s}$, and find the distance of the particle from the starting point after a further $2\,\mathrm{s}$.

(ii) Find at what time the particle's velocity is zero.

(iii) Find the total distance travelled by the particle in the first 4 seconds.

Further practice

1   The displacement x m of a moving point A at time t seconds is given by the formula

$$x = 2t^3 - 3t^2 + 4t - 10$$

Find the velocity and the acceleration of A at the instant $t = 4$.

2   A particle is moving in a straight line and its displacement s from a fixed point O is given by the formula

$$s = 18t - 21t^2 + 4t^3$$

(i)   Find the velocity and acceleration after 4 s.

(ii) Find the distance travelled between the two times when the velocity is instantaneously zero.

(iii) Find the total distance travelled in the interval $0 \leqslant t \leqslant 4$.

3   A particle is moving in a straight line. The position s of the particle at time t is given by

$$s = 18 - 24t + 9t^2 - t^3; \qquad 0 \leqslant t \leqslant 5.$$

(i)   Find the velocity v at time t and the values of t for which $v = 0$.

(ii) Find the position of the particle at those times.

(iii) Find the total distance travelled by the particle in the interval $0 \leqslant t \leqslant 5$.

4   A particle moves along a straight line so that after t seconds, $t \geqslant 0$, its displacement s from the origin O on the line is given by

$$s = (t - 1)(t - 2)^2$$

(i)   Find the velocity and acceleration of the particle on each occasion that it passes the origin.

(ii) Find the distances of the particle from O each time the velocity is 0.

(iii) Find the acceleration when the velocity is $5\,\mathrm{m\,s^{-1}}$.

5   A particle starts from rest at O and moves along a straight line. After t seconds its velocity is $v\,\mathrm{m\,s^{-1}}$, where $v = t^2 - t^3$.

(i)   Show that the particle is momentarily at rest after 1 s and find its distance from O at this time.

(ii) Find the maximum velocity of the particle in the first second of the motion.

6   A particle starts from rest at a point 5 m from O and moves in a straight line away from O with velocity $v\,\mathrm{m\,s^{-1}}$ at time t s given by $v = 3t - \dfrac{1}{12}t^2$.

(i)   Find its acceleration and distance from O, each in terms of t.

(ii) Find the time at which it begins to return, and the time at which it again reaches its starting point.

7   The velocity of a particle is given by $v = 5t - t^3$. Find the maximum displacement of the particle.

8   During braking, the speed of a car is given by $v = 12 - 3t^2$ until it stops moving. Find the distance travelled from the time that the braking starts.

9   The velocity of a sprinter at the start of a race is given by

$$v = 8t - 2t^2; \qquad 0 \leqslant t \leqslant 2$$
$$v = 8; \qquad\qquad t \geqslant 2.$$

(i)   Find the acceleration of the sprinter at $t = 2$. Hence write down the maximum speed.

(ii) How far does the sprinter run in the first 2 seconds?

(iii) How long does the sprinter take to run 100 metres?

10 Starting from rest at O, a particle travelling in a straight line is subject to an acceleration a m s^{-2} given by $a = 15 - 0.75t$.

(i) Find the particle's velocity after 15 seconds.

(ii) Find the distance travelled by the particle after 15 seconds.

(iii) At what time is the particle's maximum speed reached and what is that maximum speed?

11 A particle moves in a straight line with acceleration $a = 6t - 2$. If the particle has an initial velocity of 3 m s^{-1}, find the distance travelled by the particle in the first second of its motion.

12 The acceleration a m s^{-2}, of a car t seconds after starting from rest is

$$a = 2 + 0.8t - 0.1t^2$$

until $a = 0$. After that time the speed remains constant.

(i) Find the car's maximum acceleration.

(ii) Find the time taken by the car to attain the greatest speed.

(iii) Find the greatest speed attained.

(iv) Find the distance travelled by the car in the first 20 seconds.

13 A particle moves in a straight line such that its velocity v at time t is given by

$$v = 18 - 27t + 10t^2 - t^3; \qquad 0 \leqslant t \leqslant 5.$$

(i) Calculate the acceleration when $t = 2$.

(ii) Find an expression in terms of t for the displacement of the particle. Hence find the displacement of the particle when $t = 5$.

(iii) Explain what happens to the motion of the particle between $t = 1$ and $t = 3$.

(iv) Find the total distance travelled in the interval $0 \leqslant t \leqslant 5$.

Past exam questions

1 A particle P travels from a point O along a straight line and comes to instantaneous rest at a point A. The velocity of P at time t s after leaving O is v m s^{-1}, where $v = 0.027(10t^2 - t^3)$. Find

(i) the distance OA, [4]

(ii) the maximum velocity of P while moving from O to A. [3]

Cambridge International AS and A Level Mathematics 9709 Paper 43 Q3 June 2012

2 A vehicle starts from rest at a point O and moves in a straight line. Its speed v m s^{-1} at time t seconds after leaving O is defined as follows.

For $0 \leqslant t \leqslant 60$, $\quad v = k_1 t - 0.005t^2$,

for $t \geqslant 60$, $\qquad v = \dfrac{k_2}{\sqrt{t}}$.

The distance travelled by the vehicle during the first 60 s is 540 m.

(i) Find the value of the constant k_1 and show that $k_2 = 12\sqrt{60}$. [5]

(ii) Find an expression in terms of t for the total distance travelled when $t \geqslant 60$. [2]

(iii) Find the speed of the vehicle when it has travelled a total distance of 1260 m. [3]

Cambridge International AS and A Level Mathematics 9709 Paper 43 Q7 November 2013

3 A particle P moves in a straight line. At time t seconds after starting from rest at the point O on the line, the acceleration of P is a m s^{-2}, where $a = 0.075t^2 - 1.5t + 5$.

(i) Find an expression for the displacement of P from O in terms of t. [4]

(ii) Hence find the time taken for P to return to the point O. [3]

Cambridge International AS and A Level Mathematics 9709 Paper 42 Q4 June 2015

4 A particle P moves in a straight line, starting from a point O. The velocity of P, measured in m s^{-1}, at time t s after leaving O is given by

$$v = 0.6t - 0.03t^2.$$

   (i) Verify that, when $t = 5$, the particle is 6.25 m from O. Find the acceleration of the particle at this time.   [4]

   (ii) Find the values of t at which the particle is travelling at half of its maximum velocity.   [6]

Cambridge International AS and A Level Mathematics 9709 Paper 41 Q6 November 2015

5 Two cyclists P and Q travel along a straight road ABC, starting simultaneously at A and arriving simultaneously at C. Both cyclists pass through B 400 s after leaving A. Cyclist P starts with speed 3 m s^{-1} and increases this speed with constant acceleration 0.005 m s^{-2} until he reaches B.

   (i) Show that the distance AB is 1600 m and find P's speed at B.   [3]

Cyclist Q travels from A to B with speed v m s^{-1} at time t seconds after leaving A, where

$$v = 0.04t - 0.0001t^2 + k,$$

and k is a constant.

   (ii) Find the value of k and the maximum speed of Q before he has reached B.   [6]

Cyclist P travels from B to C, a distance of 1400 m, at the speed he had reached at B. Cyclist Q travels from B to C with constant acceleration a m s^{-2}.

   (iii) Find the time taken for the cyclists to travel from B to C and find the value of a.   [4]

Cambridge International AS and A Level Mathematics 9709 Paper 41 Q7 June 2014

6

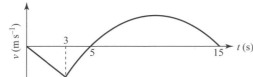

The diagram shows the velocity–time graph for the motion of a particle P which moves on a straight line BAC. It starts at A and travels to B taking 5 s. It then reverses direction and travels from B to C taking 10 s. For the first 3 s of P's motion its acceleration is constant. For the remaining 12 s the velocity of P is v m s^{-1} at time t s after leaving A, where

$$v = -0.2t^2 + 4t - 15 \text{ for } 3 \leqslant t \leqslant 15.$$

   (i) Find the value of v when $t = 3$ and the magnitude of the acceleration of P for the first 3 s of its motion.   [3]

   (ii) Find the maximum velocity of P while it is moving from B to C.   [3]

   (iii) Find the average speed of P

      (a) while moving from A to B,

      (b) for the whole journey.   [6]

Cambridge International AS and A Level Mathematics 9709 Paper 42 Q7 November 2014

7 A particle travels in a straight line from a point P to a point Q. Its velocity t seconds after leaving P is v m s^{-1}, where $v = 4t - \frac{1}{16}t^3$. The distance PQ is 64 m.

   (i) Find the time taken for the particle to travel from P to Q.   [5]

   (ii) Find the set of values of t for which the acceleration of the particle is positive.   [4]

Cambridge International AS and A Level Mathematics 9709 Paper 41 Q6 June 2011

STRETCH AND CHALLENGE

1 Two sprinters are having a race over 100 metres. Their accelerations in $m\,s^{-2}$ are as follows:

Sprinter A	Sprinter B
$a = 5.4t - 1.8t^2$; $0 \leqslant t \leqslant 3$	$a = 3.24t - 0.81t^2$; $0 \leqslant t \leqslant 4$
$a = 0$; $t > 3$	$a = 0$; $t > 4$

(i) Find the greatest speed of each sprinter.

(ii) Find the distance run by each sprinter while reaching their greatest speed.

(iii) How long does each take to finish the race?

(iv) Who wins the race, by what time margin and by what distance?

2 A particle is moving in a straight line. Starting from rest, it has an acceleration given by

$a = 1 - 0.2t$; $0 \leqslant t \leqslant 4$

$a = 0.2$; $4 \leqslant t \leqslant 10$

(i) Find the speed of the particle when $t = 4$.

(ii) Find the distance travelled in the interval $0 \leqslant t \leqslant 4$.

(iii) Find the total distance travelled.

3 The acceleration of a car t seconds after starting from rest is

$$a = \frac{50 + 5t - t^2}{25}\,m\,s^{-2}$$

until the instant $a = 0$. After this instant the speed of the car remains constant. Find

(i) the maximum acceleration

(ii) the time taken to attain the greatest speed

(iii) the greatest speed attained.

4 The distance of a particle from a point O at time t is given by

$s = 5 \sin 2t - 12 \cos 2t$

(i) Find expressions for the velocity v and acceleration a in terms of t.

(ii) Show that a and s satisfy the equation $a = -4s$.

(iii) Show that v and s satisfy the equation $v^2 = 4(169 - s^2)$.

8 A model for friction

1 Each diagram shows a block of mass 5 kg resting on a rough horizontal surface. The block is being pulled by an inextensible string with tension T. Given that the block is on the point of sliding in each case, find:

(a) the normal reaction R between the block and the surface

(b) the coefficient of friction μ between the block and the surface.

(i)

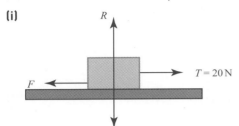

(ii)

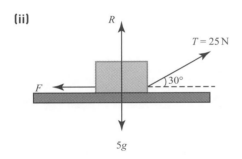

(iii)

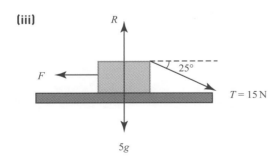

2 In each of the following situations blocks of given masses are attached to the ends of light inextensible strings going over light smooth pulleys. The coefficient of friction μ between the surface and the block that is in contact with it is also given. In each case

(a) find the acceleration

(b) find the tension in each string

(c) find the magnitude of the frictional force.

(i)

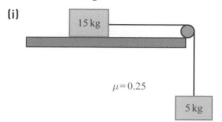

(ii)

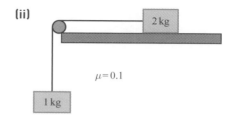

(iii)

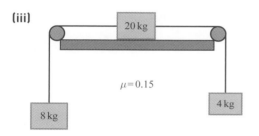

(iv)

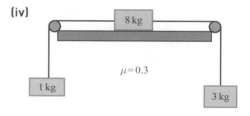

(v)

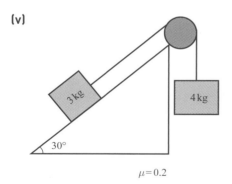

3   A car travelling at $15\,\mathrm{m\,s^{-1}}$ on a straight horizontal road skids to a halt in a distance of $30\,\mathrm{m}$.

   (i)   Find its deceleration.

   (ii) Find the coefficient of friction between the road and the car.

   (iii) Find the stopping distance from a speed of $25\,\mathrm{m\,s^{-1}}$.

4   A block of mass $5\,\mathrm{kg}$ lies on a rough horizontal table. The coefficient of friction between the table and the block is 0.2. The block is attached by a light inextensible string passing over a smooth pulley at the edge of the table to a mass of $3\,\mathrm{kg}$ hanging freely. The $5\,\mathrm{kg}$ mass is $2\,\mathrm{m}$ from the pulley and the $3\,\mathrm{kg}$ mass is $1.5\,\mathrm{m}$ from the floor.

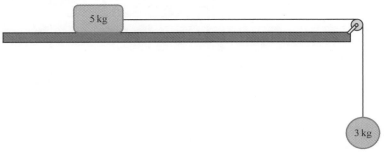

The system is released from rest. Find

   (i)   the acceleration of the system

   (ii) the time for the $3\,\mathrm{kg}$ mass to reach the floor

   (iii) the velocity with which the $5\,\mathrm{kg}$ mass hits the pulley.

5 A box of mass 10 kg is being pushed along uniform rough ground by means of a downward force of 25 N at 60° to the vertical as shown in the figure.

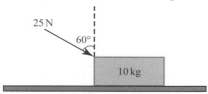

The box is initially at rest and is travelling at 0.5 m s⁻¹ after it has slid 4 m.

(i) Find the frictional force.

(ii) Find the coefficient of friction.

When the box is moving at 0.5 m s⁻¹, the force is removed.

(iii) From the point where the force is removed, how far does the box slide before coming to rest?

(iv) If the force had been 25 N upwards at 60° to the vertical, would the box have been travelling at the same speed, or faster or slower, after sliding for 4 m?

6 A stone is released from rest on a rough inclined plane, making an angle of 30° to the horizontal, and slides downhill for one metre in one second. Find the coefficient of friction between the stone and the plane.

7   The figure shows a mass of 25 kg on a slope making an angle of 30° with the horizontal. The mass is being pulled by a rope making an angle of 15° with the slope. The tension in the rope is T. The coefficient of friction between the mass and the slope is 0.3.

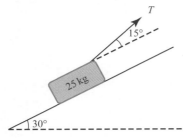

Find T, if

(i)   the mass is about to move down the slope

(ii) the mass is about to move up the slope

(iii) the mass is accelerating at $2\,\mathrm{m\,s^{-2}}$ up the slope.

8   A block of mass 25 kg is at rest on a plane inclined at 15° to the horizontal. A force acts on the block in a direction parallel to the line of greatest slope of the plane. The coefficient of friction between the block and the plane is 0.35.

Find the least magnitude of the force necessary to move the block

(i)   given that the force acts up the plane

(ii) given instead that the force acts down the plane.

Further practice

1   A stone is sliding in a straight line across a horizontal ice rink. Given that the initial speed of the stone is $5\,\mathrm{m\,s^{-1}}$ and that it slides $20\,\mathrm{m}$ before coming to rest, calculate the coefficient of friction between the stone and the ice.

2   A mass of $3\,\mathrm{kg}$ is pulled along a horizontal floor by means of a force of $50\,\mathrm{N}$ acting at $15°$ above the horizontal.

   (i)   If the coefficient of friction μ is equal to 0.3, find the acceleration of the mass.

   (ii) What would the acceleration of the mass be if the force acted downwards at $15°$ to the horizontal?

3   A child at a water sports centre is sliding down a chute which is $12\,\mathrm{m}$ long and inclined at an angle of $30°$ to the horizontal. If the coefficient of friction between the chute and the child is 0.2, find the child's acceleration down the chute and speed on leaving the chute.

4   Find the least force, acting in a direction parallel to the slope, that will move a block of mass $75\,\mathrm{kg}$ up a rough plane inclined at $20°$ to the horizontal when the coefficient of friction between the block and the plane is 0.6.

5   A block of mass $M\,\mathrm{kg}$ is lying on a rough inclined plane at an angle α to the horizontal. It is connected by a light inextensible string, which passes over a smooth pulley at the top of the plane, to a mass $m\,\mathrm{kg}$ which is hanging freely. The coefficient of friction between the block and the plane is μ.

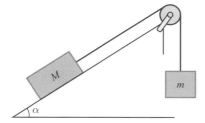

   When the system is free to move, find the acceleration and the tension in the string if

   (i)   the hanging mass descends

   (ii) the hanging mass ascends.

6   A block of mass $25\,\mathrm{kg}$ rests on a rough plane inclined at an angle θ to the horizontal with $\sin\theta = 0.35$. The coefficient of friction between the block and the plane is 0.2.

   (i)   Find the acceleration of the block when a force of $40\,\mathrm{N}$ acts on it. The force is parallel to the plane and down the plane.

   (ii) What force acting parallel to the plane would be required to give an equal acceleration up the plane?

7   A block of mass $1.2\,\mathrm{kg}$ is pulled across a horizontal surface by a force of $10\,\mathrm{N}$ inclined at an angle of $60°$ to the vertical. The coefficient of friction between the block and the surface is 0.3.

   (i)   Calculate the vertical component of the force exerted by the surface on the block.

   (ii) Calculate the acceleration of the block.

   The $10\,\mathrm{N}$ force is removed when the speed of the block is $3\,\mathrm{m\,s^{-1}}$.

   (iii) Calculate the time taken for the block to decelerate from a speed of $3\,\mathrm{m\,s^{-1}}$ to rest.

8   A block of mass $10\,\mathrm{kg}$ is placed on a rough plane inclined at $20°$ to the horizontal. A force P of magnitude $25\,\mathrm{N}$ acting parallel to the plane is just enough to prevent the block from sliding down the plane.

   (i)   Find the coefficient of friction μ between the block and the plane.

   (ii) P is now increased until the block is about to slide up the plane. Find the value of P.

   (iii) P is now increased to $100\,\mathrm{N}$. What is the acceleration of the block up the plane?

9 Two particles A and B of masses 10 kg and 4 kg, respectively, are attached to the ends of a light inextensible string which passes over a smooth pulley at C at the top of a fixed rough plane inclined at an angle of 50° to the horizontal.

The particles are at rest with the 10 kg mass in contact with the plane and the 4 kg mass hanging freely.

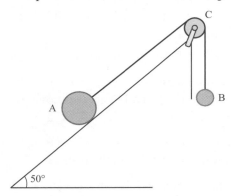

Given that A is in limiting equilibrium and on the point of moving down the plane, find μ, the coefficient of friction.

10 A box of mass 50 kg is at rest on rough horizontal ground. A force of magnitude 200 N acts upwards on the box at an angle θ to the vertical, where $\tan\theta = 0.75$.

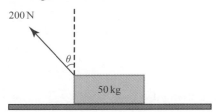

The box is about to slip. Find μ, the coefficient of friction between the box and the ground.

11 A force of $3X$ is applied to a block of mass m which is lying on a rough plane inclined at an angle α to the horizontal, in a direction parallel to the plane. This causes the mass to be on the point of moving up the plane.

A force X up the plane is applied to the same block and is just sufficient to prevent the block from sliding down the plane.

If μ is the coefficient of friction between the block and the plane, show that $\mu = 0.5 \tan\alpha$.

12 A particle of mass 2 kg is projected up an inclined plane, making an angle of 20° with the horizontal, with a speed of $6\,\mathrm{m\,s^{-1}}$. The particle comes to rest after 4 m.

(i) Find the deceleration of the particle.

(ii) Find the frictional force F and the normal reaction R, and hence deduce the coefficient of friction between the particle and the plane.

The particle then starts to move down the plane with acceleration $a\,\mathrm{m\,s^{-2}}$.

(iii) Find a and the speed of the particle as it passes its starting point.

13 Particles A and B are attached to opposite ends of a light inextensible string. Particle A, of mass m kg, is at rest on a rough horizontal table. The string passes over a smooth pulley fixed at the edge of the table. Particle B of mass 2 kg hangs vertically below the pulley. The coefficient of friction between particle A and the table is 0.45. Particle A is on the point of slipping.

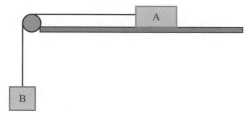

  (i) Find T, the tension in the string.

  (ii) Find m.

A particle of mass 0.5 kg is now attached to B and the system starts to move.

  (iii) Find the tension in the string.

14 A box of mass 20 kg rests on a rough horizontal surface. The coefficient of friction between the box and the surface is 0.25. A light inextensible string is attached to the box in order to pull it. T is the tension in the string.

  (i) Find the minimum value of T for the box to move when

    (a) the string is horizontal

    (b) the string makes an angle of 20° above the horizontal

    (c) the string makes an angle of 20° below the horizontal.

  (ii) If the string is pulled at 30° above the horizontal, what value of T would make the box move with an acceleration of $2\,\mathrm{m\,s^{-2}}$?

15 A box of mass 12 kg is held in equilibrium on a rough horizontal table by two light inextensible strings. The strings make angles of 60° and 30° with the horizontal on either side of the box and the tensions in the strings are 80 N and 50 N respectively.

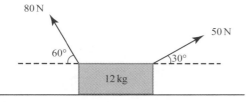

The box is on the point of slipping to the right. Find the coefficient of friction between the box and the table.

Past exam questions

1   A block of mass 11 kg is at rest on a rough plane inclined at 30° to the horizontal. A force acts on the block in a direction up the plane parallel to a line of greatest slope. When the magnitude of the force is $2X$ N the block is on the point of sliding down the plane, and when the magnitude of the force is $9X$ N the block is on the point of sliding up the plane. Find

   (i)   the value of X, [3]

   (ii) the coefficient of friction between the block and the plane. [4]

Cambridge International AS and A Level Mathematics 9709 Paper 41 Q4 June 2011

2   A string is attached to a block of weight 30 N, which is in contact with a rough horizontal plane. When the string is horizontal and the tension in it is 24 N, the block is in limiting equilibrium.

   (i)   Find the coefficient of friction between the block and the plane. [2]

The block is now in motion and the string is at an angle of 30° upwards from the plane. The tension in the string is 25 N.

   (ii) Find the acceleration of the block. [4]

Cambridge International AS and A Level Mathematics 9709 Paper 42 Q1 June 2013

3   A particle moves up a line of greatest slope of a rough plane inclined at an angle α to the horizontal, where $\sin \alpha = 0.28$. The coefficient of friction between the particle and the plane is $\frac{1}{3}$.

   (i)   Show that the acceleration of the particle is $-6 \, \text{m s}^{-2}$. [3]

   (ii) Given that the particle's initial speed is $5.4 \, \text{m s}^{-1}$, find the distance that the particle travels up the plane. [2]

Cambridge International AS and A Level Mathematics 9709 Paper 43 Q1 November 2013

4   ABC is a line of greatest slope of a plane inclined at an angle α to the horizontal, where $\sin \alpha = 0.28$ and $\cos \alpha = 0.96$. The point A is at the top of the plane, the point C is at the bottom of the plane and the length of AC is 5 m. The part of the plane above the level of B is smooth and the part below the level of B is rough. A particle P is released from rest at A and reaches C with a speed of $2 \, \text{m s}^{-1}$. The coefficient of friction between P and the part of the plane below B is 0.5. Find

   (i)   the acceleration of P while moving

      (a) from A to B,

      (b) from B to C, [3]

   (ii) the distance AB, [3]

   (iii) the time taken for P to move from A to C. [3]

Cambridge International AS and A Level Mathematics 9709 Paper 42 Q6 November 2014

5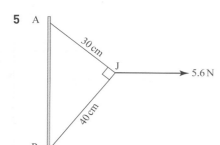

A small ring R is attached to one end of a light inextensible string of length 70 cm. A fixed rough vertical wire passes through the ring. The other end of the string is attached to a point A on the wire, vertically above R. A horizontal force of magnitude 5.6 N is applied to the point J of the string 30 cm from A and 40 cm from R. The system is in equilibrium with each of the parts AJ and JR of the string taut and angle AJR equal to 90° (see diagram).

(i) Find the tension in the part AJ of the string, and find the tension in the part JR of the string. [5]

The ring R has mass 0.2 kg and is in limiting equilibrium, on the point of moving up the wire.

(ii) Show that the coefficient of friction between R and the wire is 0.341, correct to 3 significant figures. [4]

A particle of mass m kg is attached to R and R is now in limiting equilibrium, on the point of moving down the wire.

(iii) Given that the coefficient of friction is unchanged, find the value of m. [3]

Cambridge International AS and A Level Mathematics 9709 Paper 42 Q7 June 2015

6 A rough plane is inclined at an angle of $\alpha°$ to the horizontal. A particle of mass 0.25 kg is in equilibrium on the plane. The normal reaction force acting on the particle has magnitude 2.4 N. Find

(i) the value of α, [2]

(ii) the least possible value of the coefficient of friction. [2]

Cambridge International AS and A Level Mathematics 9709 Paper 41 Q2 June 2014

7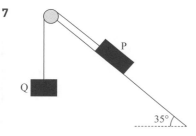

Blocks P and Q, of mass m kg and 5 kg respectively, are attached to the ends of a light inextensible string. The string passes over a small smooth pulley which is fixed at the top of a rough plane inclined at 35° to the horizontal. Block P is at rest on the plane and block Q hangs vertically below the pulley (see diagram). The coefficient of friction between block P and the plane is 0.2. Find the set of values of m for which the two blocks remain at rest. [6]

Cambridge International AS and A Level Mathematics 9709 Paper 41 Q4 November 2015

STRETCH AND CHALLENGE

1 A particle slides down a rough plane inclined at an angle θ to the horizontal such that $\sin\theta = 0.6$. The coefficient of friction between the particle and the plane is $\frac{9}{16}$.

Show that the time to descend any distance X is twice the time that would be taken if the plane was smooth.

2 A box of mass 850 kg is placed on a rough plane inclined at an angle $\alpha = \sin^{-1}\left(\frac{13}{85}\right)$. The coefficient of friction between the box and the plane is $\frac{1}{7}$. A rope is attached to the box and the direction of the rope makes an angle $\beta = \sin^{-1}\left(\frac{7}{25}\right)$ with the surface of the plane.

If the tension in the rope is T, then show that, for the box to remain in equilibrium, $109 \leqslant T \leqslant 2500$.

3 A horizontal force $4P$ applied to a mass M on a rough plane of inclination θ causes the mass to be on the point of moving up the plane.

A horizontal force P applied to the same mass on the same plane is just sufficient to prevent the mass from sliding down the plane.

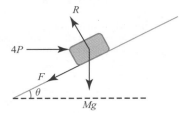 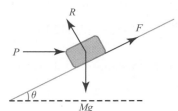

μ is the coefficient of friction between the mass and the plane.

Show that $\mu = 0.6(1 + \mu^2)\sin\theta\cos\theta$.

4 A box of weight 200 N is pulled at a steady speed across a rough horizontal surface by a rope which makes an angle α with the horizontal. The coefficient of friction between the box and the surface is 0.5. Assume that the box slides and does not tip up.

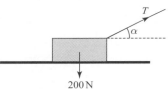

(i) Find an expression for the tension in the rope, T, for any angle α.

(ii) For what value of α is T a minimum and what is the value of that minimum?

9 Energy, work and power

9.1 Work and energy

1 A ball of mass 0.25 kg is thrown vertically upwards from a point 1 m above the ground with an initial speed of $10\,\mathrm{m\,s^{-1}}$.

   (i)   Calculate the initial kinetic energy of the ball.

   (ii) Assuming that there is no air resistance, use an energy method to find the greatest height above ground reached by the ball.

   An experiment shows that the maximum height reached is 5.5 m above ground level.

   (iii) Find the work done against the air resistance that acts on the ball as it moves.

   (iv) Assuming that the resistance force is constant, find its magnitude.

   (v)   Find the speed of the ball as it hits the ground.

2   A cyclist and her bicycle have a combined mass of 55 kg. The cyclist ascends a straight hill of constant slope making an angle θ with the horizontal with $\sin \theta = 0.25$ and of length 50 m. Starting from rest at the bottom, A, she reaches a speed of 6 m s^{-1} at the top, B. The resistance to motion is constant and has a magnitude of 50 N.

(i)   Find the gain in kinetic energy.

(ii) Find the increase in gravitational potential energy.

(iii) Find the total work done by the cyclist.

3   A ball of mass 5 kg ascends a rough slope inclined at 30° to the horizontal. The initial speed of the ball is 5 m s^{-1} and the coefficient of friction between the slope and the ball is 0.4.

(i)   When the ball has travelled 1.2 m up the slope, find

(a) the work done against friction

(b) the gain in gravitational potential energy

(c) the loss in kinetic energy

(d) the speed of the ball.

(ii) Find the total distance up the slope travelled by the ball before it comes to rest.

4   A block C of mass 15 kg lies on a smooth horizontal table. Each side of the block is connected to a small sphere by means of a light inextensible string passing over a smooth pulley as shown in the figure. Sphere A has mass 5 kg and sphere B has mass 12.5 kg. The spheres hang freely.

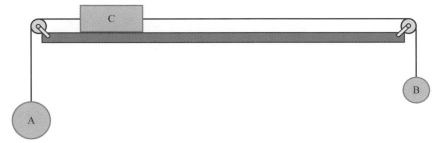

With the block at point O, the system is released from rest with both strings taut. The block reaches a speed of 0.75 m s^{-1} at point P.

(i)   Calculate the change in gravitational potential energy of the system, stating whether it is a gain or a loss.

(ii) Find the distance OP.

5 A sledge of mass 20 kg is being pulled up a slope which makes an angle of 10° to the horizontal. The sledge is being pulled by a rope which makes an angle of 15° to the slope. The sledge starts from rest and after travelling 8 m has a speed of $1.5\,\mathrm{m\,s^{-1}}$.

If resistances to motion are ignored, find

(i) the gain in gravitational potential energy

(ii) the gain in kinetic energy

(iii) the tension in the rope.

9.2 Power

1 A car of mass 1200 kg is driven along a level road against constant resistances totalling 400 N. The engine is working at a steady rate of 7.5 kW.

(i) Find the acceleration when the car's speed is $15\,\mathrm{m\,s^{-1}}$.

(ii) Find the maximum speed of the car.

2 A car of mass $900\,\text{kg}$ is travelling along a straight level road. The car's engine is developing a constant power of $20\,\text{kW}$ and there is a constant resistance of $1000\,\text{N}$.

(i) Calculate the maximum possible steady speed of the car.

(ii) Find the driving force and the acceleration of the car when the speed is $10\,\text{m}\,\text{s}^{-1}$.

When the car is travelling at $15\,\text{m}\,\text{s}^{-1}$ up a constant slope inclined at $\sin^{-1}(0.1)$ to the horizontal, the driving force is removed. Subsequently, the resistance to the motion of the car remains constant and equal to $1000\,\text{N}$.

(iii) How far up the slope does the car go before coming to rest?

3 A car maintains a constant speed of $30\,\text{m}\,\text{s}^{-1}$ on a horizontal road against constant resistances of $600\,\text{N}$.

(i) Find the driving force of the engine.

(ii) At what rate is the engine working?

4 A train of mass $3 \times 10^5\,\text{kg}$ travels along a straight level track. The resistance to motion is $1.5 \times 10^4\,\text{N}$.

(i) Find the driving force needed to produce an acceleration of $0.1\,\text{m}\,\text{s}^{-2}$.

(ii) Find the power developed by the engine when the speed of the train is $10\,\text{m}\,\text{s}^{-1}$.

(iii) Find the maximum speed attainable on the same track when the engine is working at a rate of $360\,\text{kW}$.

5 A car of mass $1250\,\text{kg}$ travelling at speed v experiences a resistance force of magnitude $25v\,\text{N}$. The car's maximum speed on a straight horizontal road is $45\,\text{m}\,\text{s}^{-1}$.

(i) Find the maximum power of the car.

(ii) Find the maximum possible acceleration of the car when it is travelling at $25\,\text{m}\,\text{s}^{-1}$.

The car starts to descend a hill on a straight road which is inclined at an angle θ to the horizontal with $\sin\theta = 0.05$.

(iii) Find the maximum possible constant speed of the car as it travels on this road down the hill.

Cambridge International AS & A Level Mathematics – Mechanics Question & Workbook

Further practice

1   A sledge slides down a straight track of length 120 m which drops down a vertical distance of 20 m. The mass of the sledge is 25 kg. The sledge's initial speed is $2\,\mathrm{m\,s^{-1}}$ and its final speed is $10\,\mathrm{m\,s^{-1}}$. The resistance force of magnitude F is constant. Use an energy method to find F.

2   A block of mass 20 kg is dragged 12 m up a slope inclined at an angle of 10° to the horizontal by a rope inclined at 15° to the slope. The tension in the rope is 100 N and the resistance to motion of the block is 50 N.

   (i)   Calculate the work done by the tension in the rope.

   (ii) Calculate the change in the potential energy of the block.

   (iii) Find the speed of the block after it has moved 12 m up the slope.

3   A mass of 5 kg is projected with a speed of $5\,\mathrm{m\,s^{-1}}$ up a smooth slope inclined at 40° to the horizontal. How far up the slope does the mass travel?

4   A particle of mass 2.5 kg slides 2 m down a rough slope making an angle of 30° with the horizontal. Starting from rest, it has a speed of $4\,\mathrm{m\,s^{-1}}$ at the end of the slope.

   (i)   Find the loss in gravitational potential energy.   (ii) Find the gain in kinetic energy.

   (iii) Calculate the work done against friction.   (iv) Find the coefficient of friction between the particle and the slope.

5   A sledge of mass 25 kg is being pulled with a force of 50 N against a resistance of 35 N. If the sledge starts from rest, what is its speed after it has travelled 25 m?

6   A truck of mass 3500 kg is moving along a straight horizontal track with a speed of $5\,\mathrm{m\,s^{-1}}$. The truck is then brought to rest by a uniform force of magnitude P N.

   (i)   Find the work done by the force.

   (ii) If the truck travels a distance of 50 m before coming to rest, find the value of P.

7   A bullet of mass 12 grams passes horizontally through a fixed board of thickness 5 cm. The speed of the bullet is reduced from $200\,\mathrm{m\,s^{-1}}$ to $120\,\mathrm{m\,s^{-1}}$ as it passes through the board. If the board exerts a constant resistive force on the bullet, find the magnitude of that force.

8   A box of mass 12 kg is placed on a rough slope inclined at an angle θ to the horizontal and $\cos\theta = 0.8$. The coefficient of friction between the box and the slope is 0.15. The box is projected up the slope from a point A with initial speed $u\,\mathrm{m\,s^{-1}}$. It travels a distance of 1.2 m up the slope before coming to rest.

   (i)   Calculate the value of u.

   As the box slides back down the slope it passes through the point of projection A and later reaches its initial speed u at a point B.

   (ii) Calculate the distance AB.

9   A small sphere of mass 0.25 kg is attached at one end, B, to a light inextensible string of length 1.75 m. The other end of the string, A, is fixed and the string can swing freely. The sphere swings with the string taut from a point where the string makes an angle of 55° with the vertical.

   (i)   Find how far down the sphere has dropped when it is at its lowest point of its swing and calculate the amount of gravitational potential energy lost at this point.

   (ii) Assuming no air resistance and that the sphere starts from rest, calculate the speed of the sphere at the lowest point of its swing.

   There is now a force opposing the motion that results in an energy loss of 0.5 J for every metre travelled by the sphere. The sphere is given an initial speed of $3\,\mathrm{m\,s^{-1}}$ and is descending with AB at 55° to the vertical.

   (iii) Calculate the speed of the sphere at the lowest point of the swing.

10 A train travels along a level track against constant resistances of 2500 N. Find the maximum speed possible if the engine works at a constant rate of 50 kW.

11 The engine of a lorry of mass 5 tonnes can develop a constant power of 75 kW.

   (i) The maximum speed on a level road is 105 km h^{-1}. Find the magnitude of the resistance to its motion.

   (ii) Find the acceleration of the lorry when it is travelling at a speed of 60 km h^{-1} against a constant resistance with the engine working at the same rate.

12 A cyclist and her bicycle have a combined mass of 75 kg.

   (i) The cyclist freewheels down a hill. Her speed increases from 3.5 m s^{-1} to 10.5 m s^{-1}. The total work done against all the resistances to motion is 1500 J. The drop in vertical height is x m. Find x.

   (ii) The cyclist then reaches a horizontal stretch of road and there is now a constant resistance to motion of 50 N. When the cyclist is developing a constant power of 250 W, find the constant speed which she can maintain.

13 The resistance to the motion of a train of total mass 200 tonnes is 16 000 N. The greatest driving force which the engine can exert is 25 000 N and the greatest power is 400 kW. The train starts from rest and moves along a level track with the greatest possible acceleration.

   (i) Show that the engine first develops its maximum power when the speed is 16 m s^{-1}.

   (ii) Show also that the speed of the train cannot exceed 25 m s^{-1}.

   (iii) Find the acceleration of the train when its speed is 20 m s^{-1}.

14 A car of mass 1250 kg travels up a straight hill inclined at 2° to the horizontal. The resistance to motion of the car is 1175 N. Find the acceleration of the car at an instant when it is moving with speed 24 m s^{-1} and the engine is working at a power of 45 kW.

15 The resistance to the motion of a car of mass 1000 kg is kv N, where v is the car's speed and k is a constant. The car ascends a hill of angle θ to the horizontal and $\sin\theta = 0.05$. The power exerted by the car's engine is 15 kW and the car has a constant speed of 20 m s^{-1}.

   (i) Show that $k = 12.5$.

The power exerted by the engine is now increased to 20 kW.

   (ii) Calculate the maximum speed of the car while ascending the hill.

16 A car of mass 1250 kg has a maximum power of 50 kW. Resistive forces have a constant magnitude of 1500 N.

   (i) Find the maximum speed of the car on level ground.

The car is now ascending a hill with inclination θ, where $\sin\theta = 0.1$.

   (ii) Calculate the maximum steady speed of the car when ascending the hill.

   (iii) Calculate the acceleration of the car when it is descending the hill at 20 m s^{-1} working at half the maximum power.

Past exam questions

1 A car of mass 700 kg is travelling along a straight horizontal road. The resistance to motion is constant and equal to 600 N.

   (i) Find the driving force of the car's engine at an instant when the acceleration is $2\,\text{m s}^{-2}$. [2]

   (ii) Given that the car's speed at this instant is $15\,\text{m s}^{-1}$, find the rate at which the car's engine is working. [2]

Cambridge International AS and A Level Mathematics 9709 Paper 41 Q1 June 2011

2

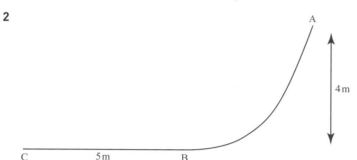

ABC is a vertical cross-section of a surface. The part of the surface containing AB is smooth and A is 4 m higher than B. The part of the surface containing BC is horizontal and the distance BC is 5 m (see diagram). A particle of mass 0.8 kg is released from rest at A and slides along ABC. Find the speed of the particle at C in each of the following cases.

   (i) The horizontal part of the surface is smooth. [3]

   (ii) The coefficient of friction between the particle and the horizontal part of the surface is 0.3. [3]

Cambridge International AS and A Level Mathematics 9709 Paper 43 Q4 November 2011

3 A lorry of mass 24 000 kg is travelling up a hill which is inclined at 3° to the horizontal. The power developed by the lorry's engine is constant, and there is a constant resistance to motion of 3200 N.

   (i) When the speed of the lorry is $25\,\text{m s}^{-1}$, its acceleration is $0.2\,\text{m s}^{-2}$. Find the power developed by the lorry's engine. [4]

   (ii) Find the steady speed at which the lorry moves up the hill if the power is 500 kW and the resistance remains 3200 N. [2]

Cambridge International AS and A Level Mathematics 9709 Paper 41 Q3 November 2015

4 A lorry of mass 16 000 kg moves on a straight hill inclined at an angle $\alpha°$ to the horizontal. The length of the hill is 500 m.

   (i) While the lorry moves from the bottom to the top of the hill at constant speed, the resisting force acting on the lorry is 800 N and the work done by the driving force is 2800 kJ. Find the value of α. [4]

   (ii) On the return journey the speed of the lorry is $20\,\text{m s}^{-1}$ at the top of the hill. While the lorry travels down the hill, the work done by the driving force is 2400 kJ and the work done against the resistance to motion is 800 kJ. Find the speed of the lorry at the bottom of the hill. [4]

Cambridge International AS and A Level Mathematics 9709 Paper 43 Q5 June 2012

5 An object of mass 12 kg slides down a line of greatest slope of a smooth plane inclined at 10° to the horizontal. The object passes through points A and B with speeds 3 m s^{-1} and 7 m s^{-1} respectively.

(i) Find the increase in kinetic energy of the object as it moves from A to B. [2]

(ii) Hence find the distance AB, assuming there is no resisting force acting on the object. [3]

The object is now pushed up the plane from B to A, with constant speed, by a horizontal force.

(iii) Find the magnitude of this force. [3]

Cambridge International AS and A Level Mathematics 9709 Paper 43 Q5 November 2012

6 A lorry of mass 12 500 kg travels along a road from A to C passing through a point B. The resistance to motion of the lorry is 4800 N for the whole journey from A to C.

(i) The section AB of the road is straight and horizontal. On this section of the road the power of the lorry's engine is constant and equal to 144 kW. The speed of the lorry at A is 16 m s^{-1} and its acceleration at B is 0.096 m s^{-2}.
Find the acceleration of the lorry at A and show that its speed at B is 24 m s^{-1}. [3]

(ii) The section BC of the road has length 500 m, is straight and inclined upwards towards C. On this section of the road the lorry's driving force is constant and equal to 5800 N. The speed of the lorry at C is 16 m s^{-1}. Find the height of C above the level of AB. [5]

Cambridge International AS and A Level Mathematics 9709 Paper 43 Q6 November 2013

7

Particles A and B, each of mass 0.3 kg, are connected by a light inextensible string. The string passes over a small smooth pulley fixed at the edge of a rough horizontal surface. Particle A hangs freely and particle B is held at rest in contact with the surface (see diagram). The coefficient of friction between B and the surface is 0.7. Particle B is released and moves on the surface without reaching the pulley.

(i) Find for the first 0.9 m of B's motion,

(a) the work done against the frictional force acting on B, [2]

(b) the loss in potential energy of the system, [1]

(c) the gain in kinetic energy of the system. [2]

At the instant when B has moved 0.9 m the string breaks. A is at a height of 0.54 m above a horizontal floor at this instant.

(ii) Find the speed with which A reaches the floor. [3]

Cambridge International AS and A Level Mathematics 9709 Paper 42 Q5 November 2014

STRETCH AND CHALLENGE

1 A railway truck runs down an incline of angle θ, such that $\sin\theta = 0.01$. At the bottom of the incline, the truck runs along a level track.

Find how far it will run on the level if the speed was a constant $8\,\mathrm{m\,s^{-1}}$ on the incline and the resistance is unchanged on the level.

2 The resistance to motion of a car is proportional to the square of its speed. The car has a mass of $1250\,\mathrm{kg}$ and can maintain a constant speed of $25\,\mathrm{m\,s^{-1}}$ when it is ascending a hill inclined at $\sin^{-1}(0.075)$ to the horizontal, with the engine working at $60\,\mathrm{kW}$.

Find the acceleration of the car when it is travelling down the same hill with the engine working at $40\,\mathrm{kW}$ at the instant when the speed is $20\,\mathrm{m\,s^{-1}}$.

3 A block of mass m rests on a rough table. The coefficient of friction between the block and the table is μ. The block is connected by a light inextensible string, which passes over a smooth pulley at the edge of the table and carries a mass M hanging vertically.

Find the velocity of the block when it has moved a distance d across the table.

4 The resistance to motion of a van of mass $M\,\mathrm{kg}$ is proportional to the square of the speed of the van. If the engine is working at $P\,\mathrm{W}$, the van can reach a maximum speed of $V\,\mathrm{m\,s^{-1}}$ up an incline making an angle θ with the horizontal.

(i) Show that the resistance when the speed is V is given by $\dfrac{P}{V} - Mg\sin\theta$.

(ii) Find the acceleration when the speed is $\dfrac{V}{2}$.

10 Momentum

1 Find the momentum of each object, assuming each of them to be travelling in a straight line.

(i) A sprinter of mass 80 kg running at $10 \, \text{m s}^{-1}$.

(ii) A tennis ball of mass 60 g moving at $250 \, \text{km h}^{-1}$.

(iii) An aeroplane of mass 5.5 tonnes cruising at $800 \, \text{km h}^{-1}$.

(iv) An ant of mass 4×10^{-3} g moving at $8 \, \text{cm s}^{-1}$.

2 A railway truck of mass 25 tonnes is shunted with speed $4 \, \text{m s}^{-1}$ towards a stationary truck of mass 20 tonnes. What is the speed of the 25 tonne truck after impact

(i) if the two trucks remain in contact

(ii) if the 20 tonne truck now moves at $4 \, \text{m s}^{-1}$?

3 A spaceship of mass 25 000 kg travelling with speed 1500 m s^{-1} docks with a space station of mass 600 000 kg travelling at 1475 m s^{-1} in the same direction.

What is the speed of the combined spaceship and space station after docking is completed?

4 A railway truck of mass 12 tonnes is moving at 8 m s^{-1} when it collides with another truck of mass 21 tonnes moving in the same direction at 4 m s^{-1}. After the collision the relative speed of the trucks is 1.5 m s^{-1}. Find

(i) the speed of each truck after the collision

(ii) the loss of kinetic energy due to the collision

(iii) the distance travelled by the lighter truck after the collision, if it is decelerated by a braking force of 3000 N.

5 A shell of mass 12 kg is fired horizontally from a gun of mass 2.5 tonnes with a speed of 350 m s^{-1}.

(i) Find the velocity of recoil of the gun.

The recoil is resisted by a constant force so that the gun moves back by 20 cm.

(ii) Find the magnitude of this force.

Further practice

1 An empty goods truck of mass M is at rest but free to move. A full goods truck of mass $3M$ travelling at $2.8\,\mathrm{m\,s^{-1}}$ runs into the first truck and is automatically coupled to it. Find the common speed after the impact.

2 A bullet of mass 25 grams is fired with a velocity of $700\,\mathrm{m\,s^{-1}}$ into a block of wood of mass 5 kg resting on a smooth horizontal surface. The bullet becomes embedded in the block. Find the common velocity of the bullet and the block.

3 A pile driver of mass 2.5 tonnes falls through a height of 3 m onto a pile of mass 0.25 tonnes.

   (i) Find the momentum of the pile driver just before impact.

   After impact the pile driver and pile move together.

   (ii) Find the common speed.

4 A bullet of mass 30 grams is fired from a gun with horizontal velocity of $520\,\mathrm{m\,s^{-1}}$. The gun has a mass of 2.2 kg. Find the initial speed of recoil of the gun.

5 A bullet is fired horizontally at $600\,\mathrm{m\,s^{-1}}$ into a block of wood of mass 0.350 kg which is resting on a smooth horizontal surface. The bullet becomes embedded in the block which starts to move with a speed of $30\,\mathrm{m\,s^{-1}}$. Find the mass of the bullet.

6 A ball A of mass 0.35 kg is moving at $4\,\mathrm{m\,s^{-1}}$ when it collides directly with a stationary ball B of mass 0.15 kg. After the collision, A continues to move in the same direction but with a reduced speed of $2.5\,\mathrm{m\,s^{-1}}$.

   (i) Find the speed of B after the collision.

   (ii) Find the loss in kinetic energy due to the collision.

7 Two marbles roll towards each other along a smooth horizontal groove. A has mass m and speed $5\,\mathrm{m\,s^{-1}}$. B has mass $2m$ and speed $8\,\mathrm{m\,s^{-1}}$. After the collision A's direction of motion is reversed and its speed is $2.5\,\mathrm{m\,s^{-1}}$. Find B's velocity after the collision.

8 A hammer of mass 0.75 kg knocks a nail of mass 20 grams into a piece of wood. Just before the impact the speed of the hammer is $5\,\mathrm{m\,s^{-1}}$. After the impact the hammer remains in contact with the nail.

   (i) Find the speed with which the nail begins to penetrate the wood.

   The wood offers a constant resistance of 5000 N against penetration.

   (ii) How far will the nail penetrate with each blow?

9 Two particles P and Q are moving in the same direction along a straight line starting 12 m apart at the same time; P starts from rest and moves with acceleration $1\,\mathrm{m\,s^{-2}}$ and Q moves with constant velocity of $5\,\mathrm{m\,s^{-1}}$. Find where they meet and, given that they coalesce, find their common velocity if the mass of P is double the mass of Q.

10 A railway truck of mass 15 000 kg moving along a level track with a speed of $3\,\mathrm{m\,s^{-1}}$ is struck by a truck of mass 5000 kg moving with a speed of $8\,\mathrm{m\,s^{-1}}$ in the same direction along the same track. After the collision the heavier truck has a speed of $4\,\mathrm{m\,s^{-1}}$. Calculate the velocity of the lighter truck just after the collision and find the total loss of kinetic energy.

11 A nail of mass 25 grams is driven horizontally into a fixed block of wood by a hammer of mass 750 grams. Immediately before striking the nail the hammer is moving horizontally at $10\,\mathrm{m\,s^{-1}}$. The hammer does not rebound and the nail is driven 5 cm into the block of wood.

   (i) Find the common speed of the hammer and nail just after the blow.

   (ii) Find the average resistance of the wood against penetration.

12 Particles A of mass 0.2 kg and B of mass 0.5 kg are connected by a light inextensible string passing over a smooth pulley.

(i) Find the acceleration of the system.

After moving for 1.4 s, B strikes the ground and does not rebound.

(ii) Find the time for which B remains on the ground.

(iii) Find the speed immediately it leaves the ground.

13 A particle of mass $5m$, which is at rest, explodes into two fragments, one of mass m, and the other of mass $4m$. The explosion provides the fragments with total kinetic energy E.

(i) Find the velocities of the fragments just after the explosion.

(ii) Find in terms of m and E, the magnitude of the relative velocity of the fragments just after the explosion.

14 Two particles of masses $2m$ and $3m$ are on a smooth horizontal table and are connected by a light inextensible string. Initially the string is slack and the particles are each moving with speed u along the same straight line and away from each other. When the string tightens, the particles begin to move in the same direction with speed v. Find v in terms of u and show that the loss in kinetic energy when the string tightens is $\frac{12}{5}-mu^2$.

▶ STRETCH AND CHALLENGE

1 A pile driver of mass M falls from a height h onto a pile of mass m, driving the pile a distance s into the ground and remaining in contact with the pile. Find an expression for the resistance of the ground, assumed to be constant.

2 A bullet of mass m, fired horizontally with velocity u, enters a block A of mass M which is free to move on a smooth horizontal table. The bullet leaves A with velocity $\frac{1}{2}u$ with its direction unaltered.

(i) Find the velocity imparted to A and the amount of energy lost owing to the passage of the bullet through the block.

After leaving A the bullet enters a fixed block B and is brought to rest after penetrating to a distance d.

(ii) Find the force (assumed uniform) exerted by B on the bullet.

3 Particles of mass m lie on a straight line on a smooth horizontal table. The first particle is projected towards the second particle with velocity u; it collides with and coalesces with the second particle and the combined particle then collides with and coalesces with the third particle. This process continues throughout the chain of particles. Find how many collisions are necessary before 99 per cent of the initial kinetic energy of the system is lost.

4 A body of mass 20 kg falls from a height of 15 m on a pile of mass 50 kg and drives it 0.15 m into the ground. The body remains in contact with the pile during penetration.

(i) Find the resistance of the ground, supposed constant.

(ii) Find the time during which the pile is moving.

(iii) Find the kinetic energy lost at the impact.

In another situation the body rebounds to a height of 0.05 m.

(iv) Find the initial velocity of the pile in that case.

5 Two particles A and B of masses m_A and m_B are moving with velocities u_A and u_B in directions inclined at an angle θ. The two particles collide and coalesce.

Find the velocity of the resultant particle and the angle that its direction makes with the initial direction of motion of A.

Reinforce learning and deepen understanding of the key concepts covered in the latest syllabus; an ideal course companion or homework book for use throughout the course.

» Develop and strengthen skills and knowledge with a wealth of additional exercises that perfectly supplement the Student's Book.

» Build confidence with extra practice for each lesson to ensure that a topic is thoroughly understood before moving on.

» Ensure students know what to expect with hundreds of rigorous practice and exam-style questions.

» Keep track of students' work with ready-to-go write-in exercises.

» Save time with all answers available online at:
www.hoddereducation.com/cambridgeextras.

Use with *Cambridge International AS & A Level Mathematics Mechanics*

9781510421745

For over 25 years we have been trusted by Cambridge schools around the world to provide quality support for teaching and learning. For this reason we have been selected by Cambridge Assessment International Education as an official publisher of endorsed material for their syllabuses.

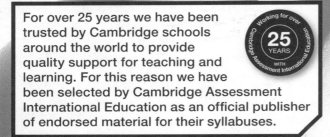

www.hoddereducation.com

ISBN 978-1-5104-2183-7

MIX
Paper from responsible sources
FSC™ C104740